Peasants, Famine and the State in Colonial Western India

Peasants, Famine and the State in Colonial Western India

David Hall-Matthews

© David Hall-Matthews 2005

All rights reserved. No reproduction, copy or transmission of this publication may be made without written permission.

No paragraph of this publication may be reproduced, copied or transmitted save with written permission or in accordance with the provisions of the Copyright, Designs and Patents Act 1988, or under the terms of any licence permitting limited copying issued by the Copyright Licensing Agency, 90 Tottenham Court Road, London W1T 4LP.

Any person who does any unauthorised act in relation to this publication may be liable to criminal prosecution and civil claims for damages.

The author has asserted his right to be identified as the author of this work in accordance with the Copyright, Designs and Patents Act 1988.

First published in 2005 by
PALGRAVE MACMILLAN
Houndmills, Basingstoke, Hampshire RG21 6XS and
175 Fifth Avenue, New York, N.Y. 10010
Companies and representatives throughout the world.

PALGRAVE MACMILLAN is the global academic imprint of the Palgrave Macmillan division of St. Martin's Press, LLC and of Palgrave Macmillan Ltd. Macmillan® is a registered trademark in the United States, United Kingdom and other countries. Palgrave is a registered trademark in the European Union and other countries.

ISBN-13: 978–1–4039–4902–8
ISBN-10: 1–4039–4902–6

This book is printed on paper suitable for recycling and made from fully managed and sustained forest sources.

A catalogue record for this book is available from the British Library.

Library of Congress Cataloging-in-Publication Data

Hall-Matthews, David (David Nicolas John) 1967–
    Peasants, Famine and the State in Colonial Western India / David Hall-Matthews.
        p. cm.
    Includes bibliographical references and index.
    ISBN 1–4039–4902–6 (cloth)
    1. Famines – India – Ahmednagar (District) 2. Peasantry – India – Ahmednagar (District) 3. Agriculture – Economic aspects – India – Ahmednagar (District) 4. Farmers – India – Ahmednagar (District) – Social conditions. 5. India – History – British occupation, 1765–1947. I. Title.

HC440.F3H35 2005
363.8'0954'79—dc22
                                                                            2004061722

10 9 8 7 6 5 4 3 2 1
14 13 12 11 10 09 08 07 06 05

Printed and bound in Great Britain by
Antony Rowe Ltd, Chippenham and Eastbourne.

For Mum, Dad and Shruti with much love and thanks

Contents

List of Tables	ix
List of Abbreviations	xi
Glossary	xii
Acknowledgements	xiv
Maps	xvi

Introduction	1
1 Landholding, Peasant Production and Rainfall	**19**
Introduction	19
The agrarian structure of Ahmednagar district	22
Distribution and fluctuation of land ownership	24
Extent of cultivation and peasant farming strategies	30
Livestock	35
Levels of production	36
Rainfall and drought	39
The impact of the famine crisis	44
Government agendas for rural development	48
Conclusion	56
2 Market Opportunities, Risks and Failures	**57**
Introduction	57
Cotton production for export	59
Knock-on benefits of the cotton boom	61
The degree of integration of Ahmednagar with external markets	64
Transport and communication infrastructure	70
Peasant poverty and rational caution	78
Free trade during the famine	82
Conclusion	88
3 Rural Moneylending, Credit Legislation and Peasant Protest	**92**
Introduction	92
The extent and causes of agricultural indebtedness	93

The nature of rural lending	95
Credit legislation and British self-criticism	97
Land transfers from peasants to moneylenders	100
The Deccan riots	104
The Deccan Agriculturists' Relief Act	107
The nexus of colonial revenue with peasant indebtedness	111
State credit schemes	114
A proposed agricultural bank in Ahmednagar	123
Conclusion	126
4 Land Revenue Rigidity, Revisions and Non-remission	**128**
Introduction	128
The importance of land revenue	129
The history of the Bombay *ryotwari* assessment	134
Difficulties in meeting the land revenue demand	138
Calculation of the revised revenue settlement	143
Revenue revisions in Ahmednagar district	150
Land revenue collection during and after the famine crisis	157
Conclusion	167
5 Peasants and Relief Labour	**168**
Introduction	168
British relief criteria	170
Relief costs	174
The nature of relief	179
Tests of eligibility for relief	184
Responses to famine policy by *ryots* and individual officers	195
Mortality	204
Conclusion	208
Conclusion	212
Notes	222
Bibliography	257
Index	264

List of Tables

1.1	Sizes of occupancies (acres) in annual *jamabandi* reports and 1884 *Ahmadnagar Gazetteer*	26
1.2	Average holding sizes by district, 1876	27
1.3	Take-ups and resignations of government land (acres), 1872–82	29
1.4	Total cultivation (acres), Ahmednagar district, 1869–83	31
1.5	Livestock totals, Ahmednagar district, 1871–81	36
1.6	Extent of cultivation (acres) and production (*maunds*) of key crops, 1871–83	37
1.7	Average yields (*maunds* and *seers*) per acre cultivated, 1874–75	38
1.8	Range of average annual rainfall measurements (inches), 1869–82	40
1.9	Rainfall figures as a percentage of monthly average, Eastern Deccan, 1876–77	43
2.1	Cotton production, Southern Division, Bombay Presidency, 1869–70	60
2.2	Annual average *jowar* prices, Ahmednagar city	64
2.3	Annual average prices for *bajri* and *jowar* in Ahmednagar, Poona and Bombay markets, 1873–74 and 1877–78 (rupees, *annas* and *pice* per *maund*)	65
2.4	Net stamp receipts (rupees), 1861–72	69
2.5	Ahmednagar stamp receipts (rupees), 1871–81	70
2.6	Cart totals in Ahmednagar district, 1876–80	74
2.7	Ahmednagar famine retail prices (rupees, *annas* and *pice* per standard *maund*)	83
3.1	Total government loans, Ahmednagar district, 1877–82 (rupees)	120
4.1	Incidence of land revenue as a percentage of produce value, Deccan, 1830–75	131
4.2	Sanctions for revenue arrears, Ahmednagar, 1873–82	142
4.3	Land revenue remission, suspension and collection, Ahmednagar district, 1867–83	158
4.4	Anticipated land revenue losses due to the 1876–78 famine, Bombay Presidency	160
5.1	Estimated scale of the 1876–78 famine	171
5.2	Famine balance sheet, Bombay Presidency	177

5.3	PWD relief wages payable in Bombay Presidency, December 1876	189
5.4	Average market rates for unskilled labour, Ahmednagar, 1869–78	189
5.5	Famine relief wages and rations, Bombay Famine Code (1885)	194
5.6	Offences and convictions, Ahmednagar district, 1874–78	199
5.7	Migration between British India and the *Nizam*'s Dominions during the 1876–78 famine	201
5.8	Monthly statement of registered deaths, Ahmednagar district, 1876–77	206
5.9	Mortality by age group and gender, Ahmednagar district, 1876–77	207
5.10	Mortality per thousand of population, Bombay Presidency, 1877	207
5.11	Rural population of Ahmednagar (village returns), 1872–81	208

List of Abbreviations

AARA	Annual Administration Reports, Ahmednagar
CD	Central Division (of Bombay Presidency)
DRCR	Deccan Riots Commission Report
EPW	Economic and Political Weekly
FCR	Famine Commission Report
GOB	Government of Bombay
GOI	Government of India
GRABP	General Report on the Administration of the Bombay Presidency
HRAD	Government of India, Home, Revenue and Agriculture Department
IESHR	Indian Economic and Social History Review
LRSB	Land Revenue and Settlements Branch
MAS	Modern Asian Studies
MSA	Maharashtra State Archives, Bombay
NAI	National Archives of India, New Delhi
ND	Northern Division (of Bombay Presidency)
NNR	Compilations of Native Newspaper Reports, Bombay Presidency
OIOC	Oriental and India Office Collections, British Library
PSS	Poona Sarvajanik Sabha
PWD	Public Works Department (Government of Bombay)
RACD	Government of India, Revenue, Agriculture and Commerce Department
RAD	Government of India, Revenue and Agriculture Department
RB	Revenue Branch
RD	Revenue Department (Government of Bombay)
SD	Southern Division (of Bombay Presidency)
SOSI	Secretary of State for India

Glossary

Anna	unit of currency; sixteen to the rupee
Annewari	system for estimating the relative yield of crops
Bagayet	irrigated land
Bajri	spiked millet, a cheap foodgrain
Bandhara	communal dam
Bania	grain trader
Bhil	member of tribal ethnic group
Brahmin	high caste Hindu, part of the traditional social elite
Candy	unit of measurement for raw cotton; 784 bales
Dacoit	bandit
Deshmukh	landholder of high social status; village representative
Dharmsala	roadside rest lodge
Dharna	fast before an adversary to demand repayment of a debt
Dufterdar	village clerk
Durbar	courtly assembly and celebration
Fellah	Egyptian peasant
Ghat	hill
Gram	type of pulse, a medium quality foodgrain
Inam	land granted at concessional revenue rates
Jagirdar	holder of an assignment of land revenue
Jamabandi	land revenue levy and collection in a given year
Jerayet	unirrigated land
Jowar	type of sorghum, a cheap foodgrain
Kharif	main annual harvest
Kharpadra	crop-destroying insect
Khyri	foodgrain disease
Koli	member of tribal ethnic group
Kulkarni	village accountant
Kunbi	peasant member of majority ethnic group in the Bombay Deccan
Lakh	one hundred thousand
Mamlatdar	revenue official responsible for a *taluka*
Manuti	interest charge calculated in grain
Maratha	member of majority ethnic group in Maharashtra, of which *Kunbis* are part
Marwari	member of ethnic group associated with moneylending
Maund	unit of measurement for grain; 80 *seers*
Mhar	member of low caste
Mofussil	interior; area away from town or headquarters

Munsif	subordinate civil judge
Nizam	prince
Nulla	drain or stream
Octroi	levy on carriage of goods
Panchayat	village council
Parsi	member of ethnic group in Bombay
Patel	village leader
Pice	units of currency; 12 to the *anna*
Pot kharab	uncultivable land
Rabi	winter harvest
Rajadharma	Hindu concept of moral rule
Ramosi	member of tribal ethnic group
Ryot	peasant
Ryotwari	land revenue system taxing landholders individually
Sarvajanik Sabha	people's association
Seer	unit of measurement for grain; just over two pounds
Sowcar	local moneylender, usually *Marwari*
Takavi	government loan for agricultural improvement
Taluka	sub-division of a district
Vadhi didhi	system for calculating interest in grain, at 50 per cent of the original loan
Vyaj	interest charge calculated in money
Zemindar	large landlord
Zemindari	land revenue system in which *zemindars* pay a fixed sum to be recouped from their tenants

Acknowledgements

I have greatly appreciated discussions and advice given by many people while researching and writing this book, including David Arnold, Jairus Banaji, Ondine Barrow, Judith Brown, Bob Currie, Lucia da Corta, Anna Lou de Havenon, Alex de Waal, Tom Downing, Jean Drèze, Mark Duffield, Nandini Gooptu, Sumit Guha, David Hardiman, John Harriss, Barbara Harriss-White, Douglas Haynes, Jaya Henry, David Keen, Jocelyn Kynch, Bishnu Mohapatra, Kenneth Parmasad, Peter Robb, Tirthankar Roy, Richard Symonds, Megan Vaughan and David Washbrook. Amrita Rangasami must take special credit for inspiring my interest in famine in India.

I am grateful for the support of colleagues at the University of Leeds notably Duncan McCargo, Ruth Pearson, Quentin Outram, Gordon Crawford, Morris Szeftel, Ray Bush, Kevin Theakston, Sarah Bracking, Felia Allum and Michael Connors. My research was generously supported by the Economic and Social Research Council and the Rajiv Gandhi Foundation. Staff at the Indian Institute Library in the Bodleian, Oriental and India Office Collections in the British Library, National Archives of India, Maharashtra State Archives, Central Secretariat Library, Ratan Tata Library, Jawaharlal Nehru University Library, Nehru Memorial Library, Gokhale Institute for Politics and Economics Library and the Ahmednagar Collector's Office Record Room have all been very helpful and efficient.

An earlier version of Chapter 2 of this book was originally published in the *Indian Economic and Social History Review*, Volume 36, number 3, © The Indian Economic and Social History Association, New Delhi, 1999. All rights reserved. Reproduced with the kind permission of the copyright holders and the publishers, Sage Publications India Private Limited, New Delhi, to whom I am obliged.

Many friends have made the task more enjoyable. I am indebted to Rieko Karatani, Hartmut Mayer, Chris Borg and Bhisham Singh. Thanks, too, to Hemendra Singh, Puvvala Prasad, Vijay and Madhu Sharma, Kirsty Milward, Raul Velasquez, Prashant Kidambi, Javed Abidi, Sanjay Barbora, Deepkanta Chowdhury, Jon Wilson, Brigid Bloom, Ben Dowson, Nicola MacNiven, Tim and Ann Chevassut, Sam Hood, Jo Morrison, Nick Alp, Chris Amis, Nicolas de Torrente, Edmund and Sarah Conybeare, Shane Doyle, Nayanika Mookherjee, Lalita Iyer,

Hamish Badenoch, Melissa Eveleigh, Isabelle Noel, Pavan Kapoor and Ramesh, Madhur, Kriti and Ashish Kapila. My family have given endless love and support, without which I could not have managed. My greatest love and gratitude is owed to Shruti, whose encouragement has been unstinting. Thank you for exhortation and inspiration, but above all for being you.

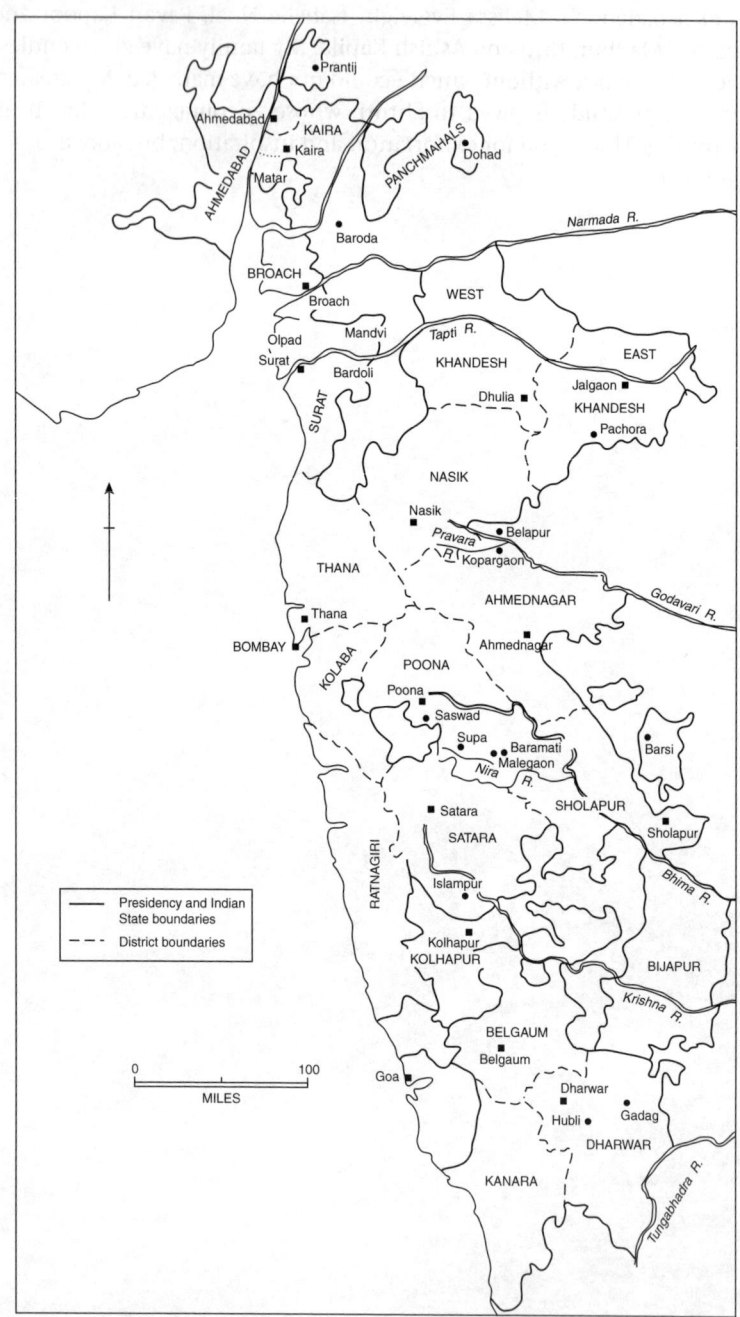

Part of Bombay Presidency showing district boundaries

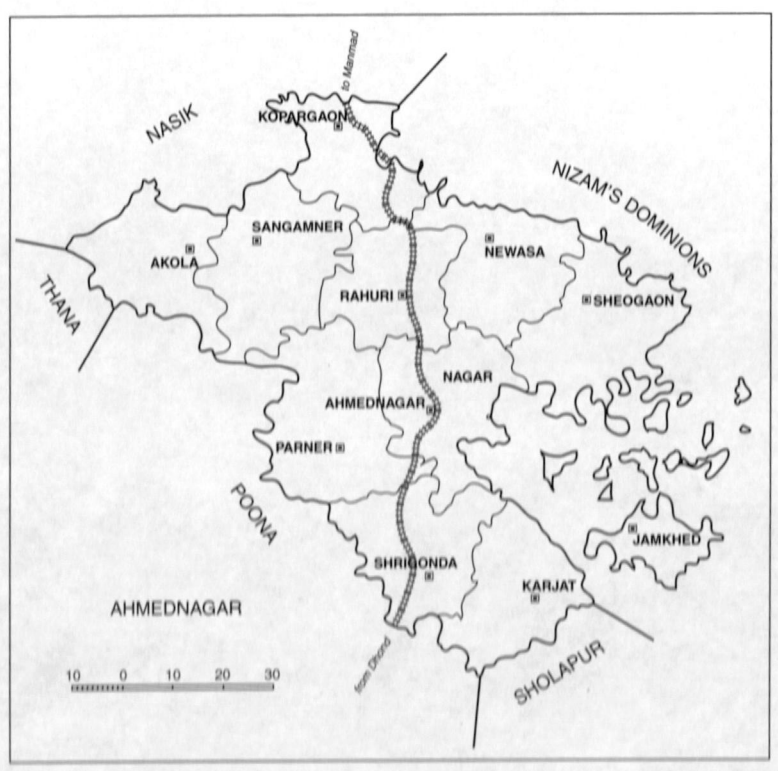

Ahmednagar District, showing divisions of *talukas*
Source: *Ahmadnagar Gazetteer*, 1884

Introduction

This book has three main objectives: first, to provide an empirical analysis of how famine developed and was responded to in an arid district over a 15-year period in colonial western India; second, through that, to develop an understanding of the relationship between the state and the peasantry, paying particular attention to the conflict between a developmentalist agenda and the food insecurity of poorer landholders; and, third, to analyse debates and decision-making at all levels of the colonial state with a view to understanding how long-term policy relating to famine prevention and relief was formulated. Thus, famine is used as a basis for evaluation of the internal workings of the colonial hierarchy and of its treatment of a significantly neglected part of its subject population. At the same time, colonial history provides an excellent context for investigation of how famines emerge and linger; of how chronic and sudden, collective and individual, endogenous and exogenous factors combine to create crises and of how crises contribute to further chronic immiseration.[1]

Famines are now widely understood not as events, but as processes, involving the economic, social and political marginalisation of parts of populations to the extent that entitlements to food and other subsistence necessities are lost. Following Amartya Sen's path-breaking assertion that famines are triggered by the collapse of exchange entitlements rather than the decline of food availability,[2] writers such as Amrita Rangasami, Alex de Waal, David Keen and Jenny Edkins have challenged his implication of a sudden accident, arguing that entitlements can also be eroded over time, often to the benefit of others.[3] Yet few local case studies exist of how particular famine processes have occurred, or of the role of states or international agencies in contributing to them or failing to prevent them. As has been argued elsewhere, India in the 1870s is a particularly useful location for such an analysis.[4] Evidence of the discussions and disputes that contributed to the development of famine policies and paradigms is more readily available than it is for recent famines and there is every reason to suppose that contemporary debates are depressingly similar. The 1880 Famine Commission Report (written to review the causes and relief of the 1876–78 famine in South India) and the Famine Codes that emanated from it set an agenda for humanitarian relief that contained temporal, spatial, social, political and financial limits that still have resonance today.

The report was also clear that the famine problem was natural, sudden and simple to explain, declaring:

> The devastating famines to which the provinces of India have from time to time been liable, are in all cases to be traced directly to the occurrence of seasons of unusual drought, the failure of the customary rainfall leading to the failure of the food crops on which the subsistence of the population depends.[5]

Addressing the same famine as the 1880 report, this book starts with the opposite assumption that the problem was human, long term and complex. Factors affecting peasants' vulnerability are explored over a period extending six years either side of the 1876–78 famine event (as conceived by the state) in Ahmednagar district, in the northeast of the Bombay Deccan. Although the significance of rain failure and exploitation by other agents, such as traders and moneylenders, is not overlooked, the emphasis is on the state's role in the famine process as experienced by Ahmednagar cultivators. In particular the book explores ways in which state policies increased or reduced these cultivators' vulnerability to economic shocks. Four categories of factors are examined: those causing chronic vulnerability, those making crisis more likely, those affecting the capacity to cope when the famine occurred and those impinging on recovery. It is not the aim here to rehearse the many debates about ideal or actual famine policy in India, only to investigate its effectiveness in Ahmednagar in this period.

This makes it necessary to consider all aspects of administration affecting the economic viability of the peasantry between 1870 and 1884. If the famine process is seen as long term, so, logically, must famine policy be, even if the Government of Bombay denied the connection between its general agricultural sector strategy and famine, and saw relief as an exceptional response to an exogenous crisis. The underlying assumptions of both agricultural policy and famine responses need to be unpacked in search of inconsistency and tested against evidence from local experience. In conceptualising a process with no clearly defined beginning or end, the importance of the post-crisis period has often been underestimated. Changes in circumstances at that point can help to identify which factors were of greatest importance in causing the crisis. Post-famine shifts in policy – either to aid recovery or in anticipation of future famines – can show how it affected government attitudes. Lack of change, on the other hand, especially in the context of a series of poor monsoons in Ahmednagar after 1878, could imply that the famine process was chronic. Further, some of the measurable effects of the crisis, including the abandonment of landholdings and losses of state land revenue, were – perhaps deliberately – not revealed in Bombay Government records for more than two years after relief had ended.

Thus, this book does not just seek to explain the 1876–78 famine in Ahmednagar. The aim is to examine in detail the conditions of peasant cultivation and the state's impacts upon them, in a specific place over a 15-year period, during the build-up to famine, the crisis and its aftermath, in order to shed light on famine process and policy. Such a particular focus needs to be set in context, as the period 1870–84 is shorter than the lifespan of several contributory factors, and the famine was far wider than a single district. In particular, peasants' difficulties in the 1870s and 1880s came in the midst of a global economic recession that hit western India hard. This followed a boom period in the 1850s and 1860s, when cultivation was dramatically extended throughout the Bombay Deccan. This book builds on Sumit Guha's argument that such expansion was economically risky, while examining the impact of the subsequent contraction of tillage on credit and commodity markets as well as household incomes.[6] As a local case study of the wider famine process, Ahmednagar offers evidence of the gradual slide into famine suffered by poorer foodgrain cultivators, but it is recognised that the same crisis affected other districts in different ways. Indeed, Ahmednagar's famine mortality was lower than that in regions where agriculture was usually more successful – perhaps because the population was more used to coping with adversity – although the effects of the crisis lasted much longer in poorer areas.

The famine process implies an incremental series of small occurrences, triggered by various natural and human agents, which make the risk of starvation more likely. It is multifaceted – a thousand cuts coming from many directions and affecting different aspects of peasants' livelihoods and strategies. Though the theoretical literature on famines has recognised the complexity of the process to some extent, Stephen Devereux has rightly identified an enduring tendency to simplify explanations of – and therefore solutions to – the famine problem.[7] Studies of famine in colonial India, like this book, have interrogated a variety of inadequacies in British rule, including its responses to famine, but have rarely gone into detail about the gradual erosion of entitlements at the local level.[8] Even a short-term local study has limitations. This book investigates the political economy of famine, more than the social history of famine causation and response. Relevant issues such as the differential experiences of men and women, of different castes, religions or ethnic groups (in particular between tribal *Bhils* and *Maratha Kunbis*) are touched on but not explored in depth, for reasons of space. Issues of health and epidemics are relevant both to famine mortality and the relationship of state to peasantry, as is crime. A substantial body of literature exists on these topics, which this book addresses only in passing.[9]

There have been long debates, considered throughout the first four chapters of this book, on how best to understand and construct late-nineteenth-century agrarian society in the Bombay Deccan, and how it would have been best to liberate its productive potential without increasing vulnerability. Writers such as Neil Charlesworth, Ravinder Kumar, Sumit Guha and Michelle McAlpin have, however, paid relatively less attention to Ahmednagar than to the other Deccan districts of Poona, Satara and Sholapur, with Jairus Banaji providing an honourable exception.[10] Chapter 1, on peasant production, shows that Ahmednagar, while not necessarily the poorest district, had a particularly undifferentiated peasantry, with fewer substantial landholders than its neighbours. Chapter 2, on trade, reveals that it was also especially isolated, with poor transport infrastructure and unintegrated markets. In combination, these factors made the district less able to benefit from the state's encouragement of commercialised markets, and kept it susceptible to the widespread depression of the 1870s. British rule did not make Deccan peasants poor, indebted or at the mercy of the season – they had been all of these for centuries. The question is rather one of how famine developed at this particular historical point, in the context of a contraction of cultivation and credit.

One of the striking features of Ahmednagar district, as discussed in detail in Chapter 2, is that the local cash crop – cotton – was grown in very small quantities, leaving most *ryots* (peasants) dependent on *jowar* (sorghum) and *bajri* (millet). Although they were cheap and bulky, these basic foodgrains were mostly sold, either in local markets or exported to districts where larger amounts of cotton were cultivated. This was because many farmers were indebted to trader-moneylenders and could not afford to store their harvest all year. It is unsurprising that foodgrain markets were less efficient and stable than those for more valuable commodities and it is argued that the role of usurious creditors as a buffer against price fluctuations was partially helpful to peasants, though it prevented them from ever realising a profit. Another uniquely unfortunate circumstance came with the implementation of land revenue increases in the district from the mid-1870s, shortly before the famine. The relationship of capricious and scanty rainfall to famine, too, in both peasant experience and colonial conception, can be highlighted in Ahmednagar, which suffered severe scarcities in 1871–72 and 1881–82 as well as in 1876–78. It was also, with Poona, the site of the Deccan Riots shortly before the famine. Chapter 3, on moneylending, considers how these riots can be analysed to shed light not only on relationships between cultivators and moneylenders, but also on the state's relationship with both and on the anticipation of impending disaster.

Every district, then, is unique and so is every famine. The 1876–78 famine in South India was one of the most devastating ever, and Ahmednagar was unusually lacking in capital and fertile land. Nonetheless, there are several aspects of the way in which the state related to the peasantry during the famine process that can be seen as typical of a wider area and of non-conflict famines in general. Agrarian structures and the poverty of factors of production were similar throughout South India and remained so for decades. The aim is not, therefore, to consider a single district crisis in isolation from its broader context but, rather, to examine Ahmednagar as a (fairly severe) example of how state policies and practice at the local level affected farmers. It is more useful, in the attempt to develop an understanding of how famines happen over time, to interrogate key aspects of the problem in local detail than to rely on aggregated colonial statistics, just as it is more meaningful to consider famine as an ongoing process throughout the period of study than to identify causes or effects of the 1876–78 crisis.

A case study of Ahmednagar also allows detailed exploration of broader themes in the context of British rule in the agricultural sector. Bombay and Madras Presidencies, where the 1876–78 famine crisis was located, were characterised by the direct *ryotwari* land revenue system which taxed landholders separately, placing peasants at the heart of government agendas for change and self-maintenance. The whole structure of colonial rule at district level revolved around the levy and collection of land revenue, the main source of state income. While the theoretical basis and intent of the Bombay *ryotwari* system remained uncertain in this period, as discussed in Chapter 4, a legal–juridical and economic language associated with the extraction of revenue directly from individual cultivators had emerged by the 1870s to replace the customary rights still understood by *ryots*. This was consistent with the imperial desire before, during and after the period of study to modernise agricultural production through sedentary tenure, capitalist accumulation and the commercialisation of land, credit and commodity markets. At the same time, the British state remained remote from peasant concerns, both procedurally and in its market-led philosophy of minimal intervention in trade or production beyond taxation.[11] Struggling Ahmednagar farmers in the early 1870s were informed, like their descendants after the re-liberalisation of the Indian economy in the 1990s and indeed like British farmers following the outbreak of foot and mouth disease in 2001, that if they could not remain independently viable they should surrender their holdings and work for wages. The logic of this choice was made harsher by the absence of state investment either in agricultural and infrastructural improvement or in alternative employment opportunities, as well as by the known food insecurity of landless labour. Famine relief, involving expensive if temporary intervention in the labour market, stood uneasily within the *laissez-faire* paradigm. It was, as a result, subject to attempts to minimise its scope – and in particular to ensure that it did not involve state participation in food markets or excessive loss of revenue – which came to a head during the 1876–78 famine.

Some historians have argued that the British state's free market philosophy and administrative distance meant that it had little direct impact on Indian rural society in this period.[12] Part of the argument in this book relates to the absence of any kind of poverty reduction strategy and the neglect of marginal sections of the population in an era before the establishment of state welfare. However, if the state made little or no difference to peasants' lives, there would be little purpose in investigating its role in the famine process. Evidence is presented throughout to show exactly how and why government policy did matter even in a remote district. Just as pro-market strategies in developing countries today often involve the active imposition of radical structural changes to the working lives and incomes of the rural poor, so did the attempt to commercialise agriculture in the mid to late-nineteenth-century Deccan. Indeed, whereas market-led policies today usually involve reduction in state subsidies for export producers, the Government of Bombay went further in forcing subsistence cultivators who had never been supported onto volatile and declining markets, through a combination of export promotion and regressive taxation.

Michelle McAlpin, against whose work on famine in western India in the nineteenth century this book is directly counterposed, argues – again as international financial institutions are wont to do today – that such short-term transitional hardship was ultimately justified by long-term improvements in the agricultural economy, resulting in a lower incidence of land revenue as well as reduced risk of famine.[13] However the very fact that the whole regional economy was in recession throughout the period of this book's study belies the teleological positivism she assumes for her longer timeframe (1860–1920). The recurrence of major famine in the region in the 1890s further suggests that things could not only get better. Despite being firmly rooted in neo-classical traditions, McAlpin's arguments ironically rest on the assertion that British economic interventions made a positive difference for the population. On the basis of careful analysis of local evidence, this book shows that her claims about market promotion and facilitation, and *takavi* (government agricultural loan) investment – both in recovery from hardship and general development – are simply not borne out. Macroeconomic policy was designed without regard for microeconomic impact and applied as if peasants were obstacles to economic development rather than its targets. Even if it were true that the long-term consequences of painful reform were beneficial for the region, it would be impossible to justify the cutting adrift of a whole generation. In fact, the almost complete absence of smallholder agriculture in Ahmednagar today makes the consequences of a *laissez-faire* sink-or-swim approach to peasant farming all too clear.

Nonetheless, the 1876–78 famine crisis covered most of southern India and was triggered by two years of extensive drought. It would therefore be folly to suggest that acts of omission or commission by the state were the only or even main cause of suffering in Ahmednagar. Mike Davis has usefully re-emphasised how provincial governments were overwhelmed by the global *El Niño* effect in a way they could neither anticipate nor control. His critique of the British is, rather, that they then used the impact of climatic crises on the population to push through their agenda of creating an unequal global capitalist system.[14] Notwithstanding the unarguable significance of simultaneous droughts in Brazil, Ethiopia and China, making famine the cause rather than the consequence of economic imperialism seems the wrong way round, and runs the danger of letting the government off the hook for its humanitarian failures. As Devereux has pointed out, 'few if any recorded famines can be blamed entirely on forces outside human influence'.[15] Periodic rain failure was a given in Ahmednagar – and the Deccan as a whole – and it is argued here that the Government of Bombay did not provide adequate safety nets precisely because it was already preoccupied with an aggressive policy of economic liberalisation. It is certainly true that some administrators were happy to see undercapitalised smallholders ruined by the famine crisis but it is argued here that this attitude was not opportunistic, as Davis infers, but had prevailed – and added to peasants' difficulties – well before the drought.

Peasants' vulnerability has often been overlooked in studies of famine because they are less severely affected than labourers or artisans, who are poorer and have fewer safety nets.[16] As in much of India, a strong weaving industry in Ahmednagar had been decimated by imports before the start of the period of study, and the continuing economic downturn hit wage labourers hardest. Most of those who died during the famine crisis in the district were landless. This does not mean that peasants did not suffer, however. The period of study shows a steady deterioration of peasant viability before, during and after the famine, including a considerable decline in cultivation in the district. Rather than the crisis polarising landholders and landless, it trapped them in a similar downward spiral.[17] As labourers died or migrated, the poorest *ryots* slid down to take their place. This was acknowledged by the Bombay Government in 1880, when they told the Secretary of State for India, 'The poorest class of cultivators succumbs to famine next after field labourers and artisans. Distress always remands the humblest of this class to the rank of field labourers.'[18] Thus a focus on the peasantry can contribute more to an understanding of the long-term process of famine. The landless were more vulnerable to fluctuations in food prices and employment opportunities, and to chronic hunger, but not to cumulative economic pressures such as debt, land revenue or falling produce prices. In 1877, Ahmednagar Collector H. E. Jacomb argued that peasants were hit hardest by the ongoing crisis: 'it will take some years before the majority of the ryots without independent means will be able to do well as an agricultural class. The labouring classes have also greatly suffered but their condition cannot be said to be so bad as that of the cultivators.'[19] This was partly because the landless had greater mobility. Although migration during the famine was a high-risk coping strategy it helped some, whereas peasants were prevented from temporary departure by the threat to strip absentee landholders of tenure for revenue default. This illustrates that peasants also had a more direct relationship with the colonial state than did the landless. Studying them can therefore offer greater insight into the impact of government policy on famine vulnerability.

The Government of Bombay itself argued that peasants could not be victims of famine, with Governor Sir Richard Temple telling the Famine Commission that neither relief nor loans need be given to landed classes because they were in no danger of starvation.[20] The commission accepted this, though they acknowledged that this made insecurity of tenure a critical issue for smaller *ryots*.[21] Such opinions reflected not only limited budgets but also the desire to draw a clear distinction between independent workers, necessarily including landholders, and paupers, for whom the British Poor Laws – the model for famine relief works – had been designed.[22] It is argued here that such lines were blurred in Ahmednagar. The optimistic attempt to make petty capitalist yeomen of struggling smallholders who, it is argued, had little economic autonomy undermined their security of tenure. This was exacerbated by guidelines intended (not wholly successfully) to exclude them from relief works. Thus peasants were unjustifiably neglected by the state. This is as relevant as ever today, with most relief targeted in the name of effectiveness despite Keen's convincing argument that other groups than the poorest can be more vulnerable to famine, particularly if they are subject to expropriation.[23] Despite the assumption that peacetime famines have now been eradicated, recent crises in southern Africa have shown that smallholders remain highly food insecure.[24] The *Bombay Gazette* insisted at the time that the success of the government's general administration as well as its famine policy should be judged by examination of the condition of those who did not take relief as well as those who did.[25]

The aim is to explore Ahmednagar peasants' structural and specific vulnerabilities within their factors of production – land, capital and labour – and within their relations to markets in both commodities and credit. Each chapter is broadly linked to one of these, attempting to tease out the forces that threatened *ryots'* livelihoods and lives. Chapter 1 looks at the productive capacity of the land in Ahmednagar, Chapter 2 the strength and accessibility of district commodity markets, Chapter 3 the role of credit, Chapter 4 the paradoxical attempt to encourage peasant accumulation through taxation and Chapter 5 conditions of labour in the specific context of famine relief works. The evolution of markets, the effects of market forces and the development – or lack of it – of factors of production are explored and deconstructed to build up a picture of the famine process in Ahmednagar. State reactions to continuity and change in these contexts, and their commentaries on them, are also emphasised to consider its role in supporting, neglecting or undermining peasant food security. The chapters are ordered according to the increasing levels of state interaction with the peasantry they involve.

The functions of the state can be usefully explored not only in the context of its interaction with the Ahmednagar peasantry but also in relation to the internal procedures of government. The process of colonial decision-making as well as the process of famine is interrogated. Famines can enable historians to glimpse the conflicting pressures on its individual agents and the varying ways with which they were dealt.

There were some within the colonial edifice who recognised the government's capacity to control economic forces which marginalised poor peasants and called for the provision of social safety nets, particularly in poorer districts and at times of crisis. However, at the initiative of its Finance Member from 1873 to 1881, Sir John Strachey, the Government of India sought to centralise decision-making and establish administrative uniformity, both in general areas such as revenue raising as well as in specific areas such as famine policy. At the same time, responsibility, especially fiscal and also for famine management, was devolved downwards to presidencies and collectors, constraining their freedom to manoeuvre.[26] The enduring paternalist imperative was offset by a drive towards high modernist bureaucratic efficiency and classical political economy.[27]

The state was not monolithic in its activities or opinions, but it was extremely hierarchical.[28] Local collectors and their assistants, who had some contact with *ryots* when touring their *talukas* (district subdivisions), had little autonomy or input into policy-making and were closely overseen to ensure that their decisions and reports were consistent with established procedures. This was done in the first instance by the divisional revenue commissioners, of whom there were variously two or three in Bombay Presidency in this period, who would report in turn to the Revenue Department of the provincial government. Any local officer inclined to make a sympathetic gesture towards the population had to seek permission by letter, which was liable to be rejected at some point in the chain of command, especially if it involved expenditure. Close knowledge of the population or changing circumstances in the district was also made less likely by a rapid turnover of staff, with only two collectors in the period writing more than one Annual Administration Report for Ahmednagar (AARA) and assistants moving on even faster. The only reference point was previous correspondence. A Revenue Department resolution was passed on virtually every letter reaching Bombay. These could vary from specific instructions and policy decisions to mere expressions of satisfaction – or otherwise – with the condition of the district and, equally importantly, its administration.

Even when the Revenue Department supported the views or suggestions of its local officers, they could come into conflict with other departments operating in the district. All local revenue officers in Ahmednagar also served the Judicial Department as magistrates but did not, curiously, have anything to do with the Survey Department, which was responsible for setting levels of land revenue on each holding, at 30-year intervals. The collector was only asked for his opinion once new rates had been fixed. As Ahmednagar's revenue revisions were assessed and implemented during and immediately after the famine crisis, this led to frequent disputes, until the Survey Department was replaced by an Agriculture Department at the end of the period of study. Fluctuating patterns of cultivation also brought collectors into conflict with the Forestry Department, which controlled acres of semi-fertile land to little effect. Attempts to improve Ahmednagar's poor infrastructure and the management of relief works led to clashes with the Public Works Department. While some of these inter-departmental arguments could be bitter, they usually revealed less about prevailing attitudes or decision-making processes than did those between officers at different levels within the Revenue Department itself. Where this book refers to the Governments of Bombay or India, the Revenue Departments are meant unless stated.

The Government of India became involved when the Government of Bombay wished to pass legislation, or passed resolutions with financial implications. Its Revenue and Agriculture Department underwent several changes of name and remit during this short period, at times including the Home and Commerce briefs. The lines of jurisdiction between the central and provincial government were tightly drawn, preventing decisions being held up in protracted disputes, but not the disputes themselves. The Government of India also frequently initiated policy discussions or sent direct instructions to Bombay. On occasion, the Secretary of State for India also intervened from London, particularly when called upon to settle disputes. On other occasions, he expressed displeasure at measures taken in Ahmednagar – particularly some revenue revisions – but was only able to review them months after they had taken place.

By focusing on correspondence and Bombay Government resolutions concerning a single district, this book is able to survey the workings of the vertical hierarchy of the colonial state, from assistant collectors in the *mofussil* (interior) up to the imperial metropolis. Examination of the nature of exchanges, records and decisions reveals more discordant discourses than shared attitudes. On occasion superior officers at every level disagreed with opinions sent to them; an assistant's suggestion was opposed by the collector, supported by the revenue commissioner, then rejected by the government. Decision-making could therefore resemble a spin of the roulette wheel, with more importance being attached to the expression of relative power than to the subject at hand. Consistency was maintained by the requirement for Annual Administration Reports to conform to pre-set sub-headings, leaving little room for local views. In 1882, for example, Collector John Elphinston was told, after suggesting that low produce prices should be taken into account when assessing revenue rates that he had 'gone beyond his province in attempting to deal with a question of this kind in a report which is intended to be strictly confined to the General Administration of the year only'.[29] Thus the colonial state was structured to make all levels accountable to those above rather than below them. District level challenges to the assumptions of the Bombay Government rarely had any impact.

It is nonetheless important to consider the variety of views expressed within such a centralised power structure, both in order to build up a picture of specific policy choices available to the government and also to explore the significance of individual agency at different levels. While the conflict between the imperatives of imperial rule and the interests of poor peasants lends itself to a structuralist approach to a question like famine process, exclusive reliance on such would generate an unsophisticated analysis. Even within rigid structures of power, paradigms can be reinforced, altered or undermined by the actions of individuals. Holders of the same office at different times have different priorities, opinions and capacities to effect change. I have argued elsewhere that the role of John Strachey was critical not only in creating enduring uniform Famine Codes at the end of this period, but also in ensuring that they prescribed the minimum possible level of state intervention and expenditure.[30] His predecessor as Finance Member, Sir Richard Temple, who was appointed Governor of Bombay from May 1877, also played a key part at that time by prioritising 'imperial discipline' over his differences of opinion with Strachey over matters of famine prevention and relief.[31] He instituted experiments designed to save money during the famine campaign – notably a lower wage on Civil Agency relief works and the suspension rather than remission of the land revenue demand – that were rejected in Madras, where he also served as Strachey's Famine Envoy.

At a lower level, William Havelock, the Revenue Commissioner of the Northern Division in the early 1870s, showed consistent concern for the condition of *ryots*, particularly on revenue matters. His successor,

E. P. Robertson, also occasionally criticised government policies but, after his promotion from the office of Collector of Khandesh, he sought to restrict comments outside his or the junior officers' jurisdiction, except over Temple's famine wage. Several Ahmednagar collectors expressed anxiety at the condition of the population throughout the period, none more so than the long-serving Henry Boswell. His retirement in September 1876, on the verge of the famine crisis he had virtually predicted, made a difference, along with Havelock's death, to the administration of the district at the time of greatest peasant vulnerability. Boswell's replacement, Jacomb, was more willing to criticise the negative impacts of policy than his own successor, T. H. Stewart, who vigorously implemented Temple's requirement to collect land revenue arrears from the famine period. However, without local experience, Jacomb was less inclined to make special relief provisions for neglected *ryots* than Boswell, or the opinionated Elphinston, who opened district relief works in 1882, may have been. On the other hand, assistant collectors – notably the long-serving Thomas Hamilton, who was not always sympathetic to poorer *ryots* – subverted the Bombay Government's guidelines by allowing them onto Civil Agency works near their homes. The apparent success of this strategy reflects the capacity of even junior officers to influence the outcomes of famine – and thus the general importance of local administrative autonomy to successful relief management. These and other examples of the difference made by personality and experience to the expression of opinions within colonial discourse are considered as a subtext in this book, along with the relative limits to the capacity of individuals to affect famine policy and process in Ahmednagar.

The focus is on colonial discussions of peasants' problems, not their own views or agency, on which much has been written in the Subaltern Studies series and elsewhere.[32] Some may find it odd that a book studying peasants at the local level fails to articulate their voice, but I hope that it does not fail to see things from their perspective. In setting out the methodologies available to the nascent Subaltern Studies school, Ranajit Guha suggested that one way to do so could be to read colonial or elite sources against the grain, and that is the intention here.[33] The question of agency is perhaps more contentious. Ahmednagar peasants had very little political or economic influence in this period and lived primarily in fear of the state. That is not to say that they never complained or protested, but they were all too rarely heard. Expedient negotiations between the state and village headmen or big peasants over the land revenue system in the earlier nineteenth century had disappeared by the 1870s.³⁴ Not only were both revenue levels and collection policy strictly predetermined in this period, local officers' powers of discretion had been significantly reduced. Close examination of colonial correspondence demonstrates the consistent lack of influence of local observations regarding the consequences of government policy. Sympathy for the population was frowned upon and peasants themselves frequently denigrated for their poverty. This is an important point, because it is argued here that new constraints on peasant autonomy under British rule were key contributory factors in the famine process. It was precisely because peasant voices were wilfully misinterpreted – especially in the context of the Deccan Riots of 1875, which were intended, in part, as a warning of impending doom – that they ended up suffering so much.

Colonial reports and correspondence also reflect an almost exclusively British view of its own administration. Though there were a couple of short-lived Indian assistant collectors in the period, records were not kept from those lower echelons of the state drawn from the *Maratha* population, *mamlatdars* (*taluka* administrators) in charge of each of Ahmednagar's 11 *talukas* and the often criticised village *kulkarnis* (village accountants). Similarly, the Famine Commission included Mahadev Wasudev Barve of Kolhapur and C. Rangacharlu of Mysore during the enquiry stage but the report was written in London in their absence. Native commentators were active outside the state, including several newspapers and the *Poona Sarvajanik Sabha* (Poona People's Association), which wrote serious reflections on key issues, such as reform of the land revenue system and provision of rural credit, in its quarterly journal and also detailed narratives of the famine and its relief. However, their opinions and evidence were never seriously considered by the Government of Bombay or the Famine Commission. The British journalist William Digby, whose own two-volume account of the famine campaign reprinted the *Sabha*'s famine narratives, argued that as they became increasingly antagonistic catalogues of complaints and criticisms, the 'suavity' of the government's response was dramatically reduced.³⁵ Nonetheless, the capacity to hold the state to account for its failures from the outside remained minimal. Both the *Sabha* and newspaper editors were drawn from proto-nationalist local elites, which were easy to dismiss as having their own agenda. The *Sabha* in particular was accused, probably falsely, of provoking both the Deccan Riots and the widespread exodus from famine relief works when wages on them were cut. On the other hand, their elitism made them at times little more sympathetic to Ahmednagar peasants or agricultural labourers than the state itself. Moreover, they made a point of not being too strongly critical of the nature of British rule. Ahmednagar Collector J. King reported typically in 1881 that in district newspapers like *Nagar Samachar*, 'the general tone of the articles published is moderate + the criticism of public measures does not exceed the limits of propriety or tend to encourage discontent or disaffection'.[36] While this might have given their views greater leverage, no exception was made to the colonial tendency of downplaying local reportage. Indeed its moderation was ensured by the Vernacular Press Act, passed in 1878. Empirical accounts of policy failure and proposals for reform are still useful as historical records with which to hold the state to account retrospectively and this book makes occasional use of such evidence – along with the report on famine management in the neighbouring *Nizam* (prince) of Hyderabad's Dominions – to support suggestions that particular colonial strategies or perceptions were inappropriate. Given their limited impact, however, the investigation into how as well as why famine could not be prevented – into the formation of policy as well as its relationship to the process of famine – necessarily relies more heavily on colonial records themselves.

The material drawn on in this work includes the Annual Administration Reports for both Ahmednagar district and Bombay Presidency. Each include statistics and commentary on revenue, trade, the extent and nature of cultivation, disease and mortality, livestock, the actions of civil and criminal courts and so on. The correspondence-based Ahmednagar reports also include more subjective discussions of issues under their formulaic sub-headings, particularly that of 'General Condition of the People'. *Jamabandi* (land revenue per year) reports, giving precise details of revenue collection, the acreages, values and yields of particular crops and, from 1874–75 onwards, the numbers and size of holdings, have also been consulted. Ian Hacking, among others, has shown the significance of such statistics-taking for maintaining control over populations,[37] while Clive Dewey has discussed both the unreliability of agricultural statistics in British India and their significance in the attempt to develop peasant agriculture.[38] Given the tension frequently created by their collection, this was an important question in Ahmednagar. Boswell, for example, complained at times that vital statistics were too arbitrarily collected to be meaningful,[39] that efforts to improve village records gave a false impression of better results year on year[40] and that demands from above for increasing volumes of forms got in the way of actual administration.[41] It was also suggested that the Survey Department's enquiries and re-measurement of holdings led to both fear of revenue rises and hope of state sanctions against *sowcars* (local moneylenders), creating a context for the Deccan Riots.[42] Issues of control are pertinent to the famine process. Aggregated agrarian statistics, like the reports, correspondence and debates that surrounded them, generated some evidence of peasant vulnerability, but did nothing to reduce it. Their collection, rather, fostered an impersonal approach and technocratic agenda for colonial rule that undermined peasant autonomy. Given their problematic nature, and the local focus of this study, more emphasis is given here to the impressions of junior officers than to the statistics themselves.

Beyond the Annual Administration, the Famine Commission and the Deccan Riots Commission Reports, more detailed Bombay Revenue and General Department correspondence is drawn on, relating to the formation of famine policy, the workings of relief, rates of pay and the suspension of the land revenue, as well as discussions of the causes of the Deccan Riots and the consequent Deccan Agriculturists' Relief Act. In addition, the Government of India's Revenue and Agriculture Department kept records of discussions of issues in Ahmednagar and Bombay Presidency. Local documents, correspondence between the two governments and secretaries of state and 'Keep-with' internal discussions within the *Takavi* and Revenue Branches have been used. These cover a wide range of subjects including state *takavi* loans, moneylending, irrigation and other infrastructural development, the state of the rural economy and above all land revenue policy and incidence. Many of these documents, especially the all-important Ahmednagar Administration Reports, take a holistic approach to the problems of the district. Questions of production, markets, indebtedness, taxation and famine were all inter-linked – and sometimes deliberately de-linked – in varying and conflicting combinations. Dividing these issues into separate chapters can only be a heuristic device, and single sources are often drawn on in several. If famine is a multifaceted process, it can shed different light according to the angle from which it is examined, while retaining its overall complexity. It cannot best be understood through a chronological narrative approach. Though relief, necessarily only relevant to the two years in which it was given, is considered separately, even that is addressed thematically in Chapter 5, rather than through a general account of the progress of events. The book investigates the process of famine throughout the period 1870–84, considering how exceptional the 1876–78 crisis was, rather than seeking its causes and effects.

Chapter 1 lays out the climatic and topographical difficulties faced in cultivating the district – in particular the variability of local rainfall. The extent to which the government was willing to treat Ahmednagar as a special case within its modernising agenda is considered alongside levels of state investment. The agrarian structure of the district is outlined to suggest that cultivators were relatively undifferentiated, with no obvious potential for the emergence of a big peasantry in the 1880s, as identified in other Deccan districts by Charlesworth.[43] The poor peasant majority was also dominated by commercial moneylenders, with no interest in re-investing in production. Patterns of landholding and cultivation are investigated, revealing a significant decline in both from the early 1870s. Chapter 2 delves further into the British desire to commercialise peasant production, considering the degree of impact of global markets on Ahmednagar dry grain farmers. Their costs and benefits are reviewed in the light of the threat of food insecurity. The district economy's levels of diversification and integration with broader markets are analysed, along with individual *ryots'* control over their commodities and access to markets, taking into account limited transport infrastructure. The particular reliance on market forces to supply food during the famine crisis is also investigated in its practical implementation. Chapter 3 examines the extent and nature of agricultural indebtedness and considers the functions of cultivators' exploitation by lenders. In the light of the Deccan Riots, for which a range of possible motivations is offered, the complicated historical relationship between *ryots*, *sowcars* and a self-critical state is explored. It is suggested that widespread usury may have served as a buffer for peasants against periodic economic shocks and the rigidity of the land revenue system, which was eroded at a critical time by plans for hostile credit legislation. The state's own rural credit system, based around *takavi* loans, is also examined. Chapter 4 follows on from this by arguing that indebtedness was largely necessitated by high and inflexible levels of taxation, exacerbated by the system of 30-year revisions. Knowledge that the revenue demand was often paid directly by *Marwaris* (moneylenders) encouraged the Government of Bombay not to curb their tax increases. Investigation of the nature and history of the *ryotwari* system and its effects on *ryots* is juxtaposed with alternative suggestions from within the colonial edifice. Particular attention is paid to controversies surrounding the refusal to remit revenue during the famine crisis and the introduction of the revised assessment in Ahmednagar in its aftermath, shedding light on tensions between the Governments of Bombay and India. Chapter 5 focuses on state and peasant responses to the famine crisis in the context of a flooded labour market. The aims, costs and conditions of relief works are considered, with a particular emphasis on the restrictions placed upon entry to them. The relative flexibility of peasants, local officers and an unprepared government in responding to the crisis is analysed in relation to migration and local works as well as mortality.

1
Landholding, Peasant Production and Rainfall

Introduction

Ahmednagar district is situated in the Bombay Deccan, due east of Bombay city, bordering Poona district to its southwest, and, in the British colonial period, the *Nizam* of Hyderabad's Dominions, in an untidy and often renegotiated border to the east. Its total area was 6666 square miles, with a population according to the 1881 census of 751,228, at an average density of 112.69 people per square mile.[1] As the average household comprised as many as nine people, the impression was of an underpopulated region with few houses, which was attributed to poor land, water supply including rainfall and other agricultural resources.[2] How did Ahmednagar peasants live and manage their land in such difficult circumstances? The Deccan Riots Commission believed that the *Kunbis*, who formed the great majority of the *ryots*, bore with 'a stubborn endurance the unkindly caprices of his climate and the hereditary burden of his debts, which would drive a more imaginative race to despair or stimulate one more intelligent to new resources'.[3] While the commission thus recognised that not all peasants' problems were naturally ordained, this interpretation, with its implication of blaming the cultivators for their stasis – and perhaps their poverty itself – reveals much about colonial attitudes to Ahmednagar district. For example, shortly after the 1876–78 famine, Survey Commissioner Colonel W. C. Anderson attacked 'pauper cultivators, without stock or means, who have in fact no business to hold land at all, but should be earning a livelihood by working for others for hire'.[4] This echoes Christine Kinealy's argument that agricultural commercialisation programmes in nineteenth-century Ireland were specifically intended to drive unproductive small peasants off the land.[5] Meanwhile the Famine Commission argued that peasants had failed to enjoy the fruits of progress through their own fault:

> It is to be expected in every forward movement in the education of a people that while the result is beneficial to the country as a whole, some classes or individuals will fail to display the qualities needed to benefit by the advantages offered, and will suffer inconvenience under the novel circumstances to which they are unable to adapt themselves.[6]

No matter that the forward movement was still to be detected in Ahmednagar, nor that inconvenience was scarcely the subject of their report. In an appendix, J. B. Peile of the Bombay Government further explained that 'The moral qualities of different races or classes of cultivators appear to have more influence on their prosperity than the pressure of rent or the quality of the soil.'[7]

Such negative perspectives underwrote the state's own unwillingness to address peasant hardship through investment. Much discourse on the condition of the district was inclined towards greater optimism than was warranted, thus further playing down peasants' difficulties and therefore any need for a response. In part due to a Whig agenda of inevitable improvement, colonial statistical records were not geared to highlight decline or the risk of famine. The Government of India acknowledged when creating the Famine Commission that the system of data collection was too subjective to be helpful for such enquiries:

> It is impossible to say, in regard to most parts of India, what proportion the produce of any one province ordinarily bears to the necessary consumption of its population, or what quantity of produce per acre constitutes an average crop; the phrase, a 12-anna crop, &c, differs with the differing personal equation of every district officer.[8]

The Government of Bombay thus had no coherent analytical framework for the understanding of peasant poverty during the period of study, nor of famine and its causation. Partly out of a desire not to intervene further in the agricultural sector, having established a legislative and taxation regime they were convinced would induce progress, the colonial hierarchy failed to explore the relationships of peasants to their land, rainfall, capital, markets, social structures and, above all, to the state itself.

Peasant cultivation in Ahmednagar was beset by a number of difficulties, some natural and some structural. Many were chronic. This chapter is concerned with how various factors impacted on farming and farming strategies, whom they affected and how they were managed – or could have been managed better. An attempt is made to see in what ways and to what extent ordinary agricultural difficulties before and after the crisis differed from those during it. This will enable consideration of whether the famine was brought about by extreme exogenous factors, or whether long-standing problems were getting worse prior to the famine, as well as of its longer-term consequences. Ideally then, it would be useful to know as much as possible about the district's cultivation patterns: how large an area was cultivated, by whom, with what crops, using what tools and labour (both human and bovine), on what quality of soil, making what profits and faced by what obstacles to production and profit, including unreliability of rainfall, diseconomies of scale and structures of exploitation that prevented reinvestment in improved techniques or technology. Plenty of aggregate information exists on most of these subjects, but it reveals little about the particular experiences of individual *ryot* households. The lowest levels for which information can be garnered from colonial records is that of the *taluka*. It is therefore hard to estimate average household production and consumption, and still harder to perceive how much they varied between households and seasons, and related to the minimum requirements for subsistence. Inevitably, data was even more aggregated when it was considered at higher levels, particularly by the Government of Bombay, which made final judgements on that basis, making their decisions insensitive even to known local variations. Nonetheless, debates between different levels of the state provide a rich source of contradictions, which reveal attempts both to assert and deny people's hardship – and also the circumstances and sometimes the means of such assertions and denials.

With an eye on the problematic and contested nature of many of the colonial sources, then, this chapter seeks first to outline the agrarian structure of the district, in order to establish the social value of production, trade and finance. In particular, the relationship of peasants to moneylenders and landlords and their consequent insecurity of tenure is outlined. It is contended that, perhaps unlike other districts nearby, elites in Ahmednagar were almost exclusively from the commercial classes, with few, if any, cultivator capitalists at this time. Further examination of land distribution into predominantly small holdings and the earlier expansion onto poorer lands leads into an analysis of the nexus between indebtedness, over-taxation and declining land ownership and cultivation well before the famine. The spatial and temporal variability of rainfall is investigated, with the argument that local rain failures were downplayed by the government but the significance of wider drought exaggerated during the famine to allow it to be ascribed as its solitary cause. Production patterns, including generally low and declining yields, livestock inputs and profitability are examined over time. The impact of the famine on long-term production is considered in the context of a particularly strong colonial desire to see signs of improvement in its aftermath. This is finally contrasted with agendas for state investment in agricultural production in Ahmednagar, including inappropriate research and inadequate irrigation, which give the impression that neither peasants nor the remote, unprofitable district as a whole were afforded serious priority by the Government of Bombay.

The agrarian structure of Ahmednagar district

The 1872 census recorded that the ratio of agriculturists to non-agriculturists in Ahmednagar district was 34.3 : 65.7 per cent, even though the former included proprietors, tenants and labourers. The unreliability of such British statistical analysis was acknowledged by the Government of Bombay, who noted, 'These figures would, however, probably be reversed were all those included in the first class who derive their support indirectly from agriculture, as, for instance, the families of farmers.'[9] In addition to all rural women and children being bizarrely labelled non-agricultural, rural creditors and tradesmen paid in kind were more legitimately excluded. The Deccan Riots Commission guessed the figure was closer to 75 per cent concerned with agriculture.[10] Judging the proportion that held no land proved equally problematic, because categories were not clearly demarcated. Many small peasants – or their families – worked on others' land as well as their own.[11] The 1872 census recorded that of the male adult population of 278,462 in the district 114,948 were cultivating proprietors and 4,716 were non-cultivating proprietors. Like the number of tenants, which was 3,687, this was a low proportion compared to the rest of the presidency. Wage (or kind) labourers totalled 22,381, a relatively high 8 per cent. These figures left 145,732 men unaccounted for, which was still probably higher than the real number not involved in agriculture,[12] especially in the absence of many non-agricultural employment opportunities, with local crafts declining from a low base during the 1870s.[13]

While labourers' incomes were most vulnerable to the vagaries of the season, lack of capital and insecurity of tenure put peasants in an uncertain position too. The Government of Bombay asserted as late as 1874 that 'A village is for Government or social purposes complete in itself, and is, so to speak, independent of the outer world.'[14] This did not mean that they were economically self-sufficient. Most grain was sold to traders and resources for cultivation purchased from external sources. The *Ahmadnagar Gazetteer* equated 'capitalists' with trading classes, a number of whom were also moneylenders. It estimated that between 50 and 75 per cent of lenders were *Marwaris*, 10 per cent *Brahmins* and the remainder local *banias* (grain traders) and *sowcars*, many of whom themselves relied on *Marwari* credit.[15] This high proportion of *Marwaris* was a direct result of British legislation in 1827 to allow the adjudication of debts in civil courts, making rural lending less risky and less dependent on the local knowledge of the previously dominant *Brahmin* lenders. This opened the door for profitable penetration of even regions as poor as Ahmednagar by immigrant creditors, whose dealings were frequently criticised by the British. The *Ahmadnagar Gazetteer* reported, 'Of all lenders the *Marwari* has the worst name. He is a bye-word for greed and the shameless and pitiless treatment of his debtor.'[16]

Estimates for the proportion of *ryots* in debt varied between two-thirds and three-quarters. The prevailing consensus was that professional trader moneylenders did not want to take over smallholdings, so much as to hold peasants in thrall via the threat of court eviction orders, and extract the entire surplus value of their production. *Marwaris* had little interest in the land and rarely permitted or underwrote farm improvements, such as well building, which might have enabled peasants to pay off their debts.[17] On the contrary, civil courts entitled them to undermine existing production by seizing agricultural utensils and cattle in part payment of debts. However, if *ryots* fell too far behind, lenders were not averse to executing mortgage bonds to cut their losses in Ahmednagar.[18] Significant numbers of holdings were resigned or transferred to non-cultivators in the early 1870s[19] and more during and after the famine, as *ryots* became decreasingly viable.[20] This was exacerbated by the Deccan Agriculturists' Relief Act of 1879, designed to restrict lenders' profits, as is seen in Chapter 3.

Many British officers argued therefore that, far from allowing enrichment through the accumulation of land value, the right of land transfer increased poverty.[21] William Pedder of the Government of Bombay pointed out further that the social status attached to land made economic land values meaningless and the right an unwanted one for landholders. He criticised the basis of the law on English experience, where other employment opportunities were plentiful, warning, 'In India it is not so. The landowner who has lost his estate sinks into abject poverty, embittered by the memory of the position he has lost, and, if a man of energy or influence, becomes politically dangerous.'[22] Such landed social elites as there had been in Ahmednagar had declined economically long before this period,[23] and Ravinder Kumar gives examples of *patels* (village headmen) in Parner *taluka* losing fertile lands through debts in the 1870s to become landless labourers.[24] Not all local officers were opposed to this process. Assistant Collector Thomas Hamilton wrote in 1879 that

> the changes that are now going on silently are such that the future condition of the people can only be guessed at. I allude to the gradual transference of land from the cultivators to the traders. My own impression is that the notorious improvidence of the cultivating classes is a complete bar to any deprecation of the change in question – the land is passing to those who are able to cultivate it and to withstand the effects of even a couple of successive bad seasons.[25]

Hamilton recognised, however, that this process also increased vulnerability, adding, 'It may be fairly anticipated that the condition of the merely cultivating classes will be lowered.'[26]

Cultivators' security and opportunities to profit were further eroded in this period by legislation regulating the relationships between landlords – many of whom were former *banias* – and tenants.[27] Although sub-letting was not common in Ahmednagar, concern was expressed in 1871 that, under the Bombay Act of 1865, landlords had the right to evict at any time for non-payment of rent. This was charged as a proportion of profits, in good seasons up to three times what landlords had paid in fixed land revenue. Collectors were obliged to assist landlords in collecting rent, although many refused to do so for sums higher than the land revenue.[28] Yet on the verge of the famine, the government passed a resolution allowing its officers to help landlords who could demonstrate that customary rents were higher than those set by the Survey Department.[29]

Distribution and fluctuation of land ownership

Lack of independent capital was critical to the vulnerability and low productivity of the Ahmednagar peasantry, both because it created a poverty trap in the long term and also because it reduced security of tenure to an increasing degree during the 1870s and 1880s. Land itself, however, was the key factor of production, and its use and distribution requires detailed consideration, in particular of the scale and pattern of individual cultivation. Holding sizes were not recorded until 1874–75, so it is impossible to determine fluctuations before the famine event. Moreover, there are discrepancies between the average occupancy size of just under 40 acres recorded in *jamabandi* reports and that estimated in other sources. The Deccan Riots Commission Report, for example, reckoned that *ryots* cultivated between three and nine acres each – a significantly low figure given the poor soil quality of the district.[30] There are three possible explanations for this divergence. The Commission's estimate is likely to have reflected acres of cultivation, rather than ownership – which were significantly higher, as is shown shortly – and further, to have interpreted the median size as a more meaningful average than the mean. Both of these suggest that their figure gives a more useful impression of farming patterns, but it is also possible that they divided the cultivated area by the adult (or male adult) population, reaching a skewed impression of the average household holding size. More disconcertingly, the 1884 *Ahmadnagar Gazetteer*, apparently calculating on the same basis as *jamabandi* reports, found many more holdings, to produce a mean size of only 15 acres. The differences are so extreme as to be impossible to explain, and undermine colonial claims of a scientific approach to statistical records. It is possible that the *Ahmadnagar Gazetteer* had a different understanding of what constituted a holding, and its data was probably collected by different people including a higher proportion of Europeans, but there is no reason to suppose that it was more accurate. The different figures are given in Table 1.1.

It would appear that fluctuations in numbers and sizes of holdings were relatively minor, except for the drop in 1879–80, when famine resignations by emigrants were belatedly acknowledged. The slight downward trend in both numbers of holdings and their average size suggests that though the smallest holders were the most likely to resign, some estates were decreasing in acreage. Study of the same figures broken down by *taluka* confirms this, in broad correlation to both seasons and land revenue increases. On the other hand, there are times at which the number of larger holdings increased, which is likely to reflect the takeover of a number of holdings by single moneylenders, out of necessity, or by acquisitive larger peasants. Leaving aside the problematic *Ahmadnagar Gazetteer* figure, the overall average size of occupancies seems large. Rather than reflecting any degree of peasant capital, though, this is

Table 1.1 Sizes of occupancies (acres) in annual *jamabandi* reports and 1884 *Ahmadnagar Gazetteer*

Acreage	1874–75	1875–76	1876–77	1877–78	1878–79	1879–80	1880–81	1881–82	1882–83	1884 Gazetteer
<5	3,709	3,712	3,499	3,804	3,786	3,784	3,841	3,760	3,862	43,404
5–10	5,009	5,012	5,059	5,206	5,275	5,235	5,095	5,075	5,071	22,723
10–20	14,229	14,462	14,457	14,709	14,933	14,296	14,250	14,234	14,703	52,079
20–50	27,453	27,271	27,268	26,908	27,473	26,272	25,446	26,256	26,990	38,812
50–100	12,820	12,770	12,914	12,681	12,708	11,962	11,940	11,897	12,177	2,995
100–200	3,488	3,514	3,587	3,393	3,413	3,319	3,275	3,290	3,434	908
200–300	332	351	401	349	327	315	315	348	327	111
300–400	93	87	94	86	66	59	57	54	63	27
400–500	25	26	21	19	17	15	17	20	23	—
500–750	12	11	10	9	11	11	11	11	10	—
750–1,000	6	6	6	5	5	5	5	4	3	48
1,000–1,500	3	3	3	3	2	3	2	3	3	—
1,500–2,000	0	0	0	0	1	0	0	0	0	—
>2,000	1	1	1	1	1	1	0	1	1	—
Total holdings	67,180	67,226	67,320	67,173	68,018	65,277	64,254	64,953	66,667	161,107
Total acres held	2,631,671	2,618,336	2,606,630	2,600,043	2,614,280	2,477,026	2,466,725	2,443,146	2,432,748	2,396,335
Mean size (acres)	39.1	38.9	38.7	38.7	38.4	37.9	38.4	37.6	36.5	14.9

Sources: Annual *jamabandi* reports – J. E. Oliphant, Acting Revenue Commissioner, Southern Division (SD), to Government of Bombay (GOB), No. 539, 24 February 1876; Maharashtra State Archives, Bombay (MSA), GOB, Revenue Department (RD) Vol. 16 of 1876, No. 62, p. 133; Robertson to GOB, No. 3654, 4 August 1877, MSA, GOB, RD, Publication No. 14593, p. 198; Robertson to GOB, No. R/626, 16 February 1878, MSA, GOB, RD, Publication No. 17193, p. 147; Robertson to GOB, No. R/336, 8 December 1879, MSA, GOB, RD, Vol. 33A of 1879, p. 543; Robertson to GOB, No. R/624, 25 February 1880, MSA, GOB, RD, Vol. 30 of 1880, p. 382; Robertson to GOB, No. R/676, 17 February 1881, MSA, GOB, RD, Vol. 28 of 1881, p. 431; Robertson to GOB, No. R/817A, 25 February 1882, MSA, GOB, RD, Publication No. 14590, p. 123; Robertson to GOB, No. R/1735, 10 April 1883, MSA, GOB, RD, Publication No. 13963, p. 119; Robertson to GOB, number erased, 14 May 1884, MSA, GOB, RD, Publication No. 14708, p. 121; *Ahmadnagar Gazetteer*, p. 244.

Table 1.2 Average holding sizes by district, Bombay Presidency, 1876

District	Holdings under 5 acres	Holdings over 1,000 acres	Average size of holdings (acres)
Ahmednagar	3,712	4	38.9
Poona	12,444	4	30.0
Kaladgi	2,957	21	37.4
Ratnagiri	57,629	10	12.2
Belgaum	9,018	19	25.3
Dharwar	6,916	11	28.4
Satara	19,933	36	24.2
Kanara	14,711	0	10.4
Sholapur	1,304	9	52.9

Source: Robertson to GOB, No. 3654, 4 August 1877, p. 32.

consistent with both low population density and the aridity of much of the soil, making farms smaller than 39 acres potentially unviable unless well located. Soil quality varied between pockets of fertile black earth on hillsides or near rivers, moderate red soil and that labelled as grey, which was in large part rock. Comparison with other districts suggests an inverse correlation between soil fertility (and therefore aggregate wealth) and average holding size, as shown in Table 1.2. Adjusting for the number of very large farms which skewed the mean in Satara, the richer Konkan districts of Dharwar, Ratnagiri, Kanara and Belgaum are seen to have smaller average holdings than the drier Deccan districts of Ahmednagar, Poona, Satara and Sholapur.

The 1884 *Ahmadnagar Gazetteer* reported that 74.6 per cent (2,750,239 acres) of land in Ahmednagar was arable, with 9.8 per cent uncultivable and 12.5 per cent reserved by the Forestry Department.[31] Of the cultivable land, 87.4 per cent was *jerayet* unirrigated land for dry cultivation, and just 2 per cent *bagayet*, watered garden land, attracting over four times higher land revenue rates. Uncultivated land amounted to 10.4 percent. However, these statistics exaggerated the natural potential of Ahmednagar district. The revision of the revenue survey re-defined the notion of cultivable land in order to make large areas liable for taxation, as is seen in Chapter 4. Although this partly reflected peasants' attempts to grow crops on ever drier, harder soil, the effect was to create an unrealistic impression of so straightforward a matter as the nature of the land itself. What land can be cultivated is a difficult question, but the official figures for the cultivable area bore little relation to annual levels of cultivation. In statistics compiled by the secretary of the 1880 Famine Commission, Sir Charles Elliott, Ahmednagar was said to have had an average of 2,433,000 acres occupied between 1873 and 1876, of which 2,152,000 were cultivated – 58.4 per cent of the total area.[32] Similarly, the area of forest reserve was extended in the 1870s as a result of re-marking, but reflected at best good intentions rather than levels of tree cover.

Essentially conservative *ryots* sought to retain their full holdings but only to cultivate them fully when seasonal prospects were favourable. Denied the chance to leave land in bad years then expand cultivation in good by the Government of Bombay (who saw such tactics as at best a regression towards unmanageable shifting cultivation and at worst tax evasion), peasants were reluctant to relinquish land they could not cultivate. It is, therefore, significant that district records show considerable fluctuation in the amount of government land surrendered and taken up during this period. Annual Administration Reports only reported the latter – itself a telling indication of the lack of concern for poorer peasants' declining viability. However, the figure given for cultivable waste land changed each year by the difference between the totals of land taken up and abandoned, making the latter possible to calculate approximately, as in Table 1.3. The figure given for land abandoned is the total waste plus land taken up for that year, minus the total waste for the previous year.

In 1878–79, the fact that as much land was taken up as resigned implied that land which had been abandoned had some value, which the British were not slow to celebrate. However, that was the exception in this period, probably reflecting resignations of better land as a result of the famine, or simply transfers to moneylenders. It was also a rare year of good rainfall. Otherwise, the steady increase in waste until 1881 demonstrates both a lack of desire to speculate in new cultivation and the insecurity of existing tenures, before and after, as well as during, the famine crisis. There was also a much larger amount of cultivable waste in Ahmednagar than other Deccan districts.[33] Long-serving Collector Henry Boswell dated the start of the local decline from the 1871–72 drought, with waste land increasing by nearly a third as a result.[34] The viability of smallholder cultivation on dry land thus decreased over a considerable time.

There were several reasons for this, notably falling foodgrain prices, as is seen in Chapter 2. As noted earlier, however, the *ryotwari* land revenue system reduced cultivators' capacity to adapt to adverse economic conditions. Up to the early years of British rule, many *Kunbis*, as well as tribal groups, cultivated different land in different years. This permitted

Table 1.3 Take-ups and resignations of government land (acres), Ahmednagar district, 1872–82

Year	Land taken up	Total waste	Land abandoned
1872–73	1,876	105,109	n/a
1873–74	3,228	108,076	6,195
1874–75	1,675	124,323	17,922
1875–76	—	148,130	23,807
1876–77	12,210	162,897	26,977
1877–78	12,965	169,483	19,551
1878–79	33,921	168,214	32,652
1879–80	6,885	223,896	62,567
1880–81	6,123	252,119	34,346
1881–82	22,187	241,291	11,359

n/a = not available.

Sources: Boswell to Havelock, No. 1645, 20 July 1874, p. 139; Boswell to Oliphant, No. 2132, 20 July 1875, Annual Administration Report, Ahmednagar (AARA) (1875), p. 494; Boswell to Havelock, No. 1952, 20 July 1876, p. 254; Jacomb to Robertson, No. A/4960, 19 July 1877, p. 47; T. H. Stewart, Acting Collector of Ahmednagar, to Robertson, No. 3195, 22–4 July 1878, AARA (1878); MSA, GOB, RD, Vol. 8 of 1878, No. 1046, p. 326; King to Robertson, No. 3140, 19–23 July 1879, AARA (1879), p. 52; King to Robertson, No. 4161, 20 July 1880, AARA (1880); MSA, GOB, RD, Vol. 10 of 1878, No. 1216, p. 73; King to Robertson, 22–5 July 1881, p. 145; J. Elphinston, Acting Collector of Ahmednagar, to Robertson, No. 5730, 20 July 1882, AARA (1882), p. 195.

them to respond to market conditions as well as the weather, by changing the extent and nature of their cultivation, and also allowed the soil to regenerate. In addition, Sumit Guha has suggested that cultivators tended to migrate frequently between the *Nizam's* Dominions and Bombay Presidency, in search of benign rule or lower taxes.[35] The *ryotwari* system permanently sedentised the peasantry, by allotting particular land to individuals. Those who resigned their holdings were unable to take up new ones elsewhere. The logic that landowners could achieve prosperity that shifting cultivators could not, including through the increase in value of their land, was scarcely borne out in a district where conditions for agriculture were so variable. Furthermore, this meant that the fluctuations in resignations and take-ups of land seen earlier could not have reflected the varying fortunes of individuals, as take-ups were not allowed by former holders. Every resignation was permanent.

Nor, given *ryots'* known unwillingness to surrender their holdings, even when they lacked the resources for cultivation, were many resignations voluntary. A large number were forfeited due to failure to pay the land revenue demand. Thus the state, as the overall landlord, increased pressure on peasants in extended periods of difficulty, such as during 1870–84. This was complicated, however, by the Government of Bombay's refusal to accept resignations during the famine from those still owing revenue. This explains why relatively little land is recorded as abandoned in Table 1.3 in 1876–78 and so much in 1879–80. Peasants who had migrated or died were recorded as holders until the figures were readjusted, minimising the impact of the famine crisis on landholding in statistical records. In 1879–80, Collector King was then able to argue that the statistical impression of continuing decline was false.[36] Revenue Commissioner E. P. Robertson agreed that only future fluctuations would reflect levels of prosperity.[37] Thus account was never taken of famine resignations, and real decline in 1879–80 was also overlooked. When abandonments far exceeded take-ups again the next year, Deputy Collector Balkrishna Deora insisted that this did not reflect levels of prosperity either, for the same reasons. Falls were solely due to belated recognition of resignations and 'the area actually cultivated in the year was larger than that of the year preceding'.[38] This was again convenient for the government, which had long been content to measure holdings rather than cultivated area for tax purposes, but King baulked at the claim of improvement, demanding statistical evidence.[39]

Extent of cultivation and peasant farming strategies

Levels of actual cultivation were always lower than ownership of designated cultivable land because of peasant strategies, both agricultural and economic. Although failure to farm assessed land could imply both seasonal and cumulative poverty, it was also partly due to fallowing. Though expensive when land revenue still had to be paid, this strategy was important to maintain soil productivity, particularly in the absence of much irrigation or manure. As seen by the need for Elliott to make his own calculations for the Famine Commission Report, the extent of cultivation was also harder to determine than the amount of land occupied. The only available data was unreliable village estimates rather than revenue records. Moreover, while cultivation records can potentially offer a more accurate insight than land ownership into the success of the agricultural economy of the whole district, their aggregation disguises the difficulties of particular *ryots*. Nor do such figures show where smallholders had surrendered their land to creditors. Notwithstanding this, as with total occupancy, a gradual decline of

Table 1.4 Total cultivation (acres), Ahmednagar district, 1869–83

Year	Acres cultivated	Change from previous year
1869–70	2,448,585	+10,995
1870–71	2,467,638	+19,053
1871–72	2,467,545	−93
1872–73	2,455,544	−12,001
1873–74	2,448,749	−6,795
1874–75	2,432,354	−16,395
1875–76	2,418,593	−13,761
1876–77	2,407,660	−10,933
1877–78	2,401,589	−6,071
1878–79	2,415,167	+13,578
1879–80	2,277,538	−137,629
1880–81	2,267,346	−10,192
1881–82	2,278,125	+10,779
1882–83	2,300,556	+22,431

Source: *Ahmadnagar Gazetteer*, p. 555.

tillage can be seen from the beginning of the 1870s through to 1881, when it bottomed out. This is shown in Table 1.4.

There was also a considerable drop caused by the famine, again appearing belatedly in the records in 1879–80. However, 66,142 acres had already gone out of cultivation before the famine started, meaning that a huge total of 200,292 acres were lost in the 1870s. To put this into perspective, the amount of land believed to be cultivable which was not tilled quadrupled in Shrigonda *taluka*, for example, in the four years after 1871–72, when cultivation was said to be at its peak.[40] Boswell suggested tentatively before the famine that contraction of agriculture in Ahmednagar may not be a wholly bad thing, because land better used as pasture had been so unprofitably sown. However, he maintained that it was also an indisputable measure of declining prosperity, warning, 'whatever the results of this decrease in the area of cultivation its cause is the growing poverty of the ryot'.[41] In retrospect, this can be interpreted as a steady slide into famine. The debate concerning causes of peasant poverty needs to be seen in this context, as some explanations not only denied links with state policies, but played down the significance of the issue. For example, Hamilton suggested that an annual decrease in the area cultivated 'is hardly to be wondered at when the capricious rainfall is taken into consideration'.[42] This was dismissed as an inadequate explanation by Boswell,[43] and the Government of Bombay raised the more pertinent point that 'The steady fall of prices which has been going on for a few years back is calculated to induce a curtailment of the area under cultivation.'[44] There was little that the government could do to increase prices, and nor was it a problem confined to poor *ryots*, although they noted that it was 'naturally most felt in districts where jowari and bajri, the two main food crops, are grown; it is least felt where exportable articles are largely raised'.[45] Local commodity markets are examined in Chapter 2.

Ahmednagar peasants were also reported to be slow to improve cultivating efficiency by new technologies or methods, even when they were available.[46] Peasant farming strategies are of key importance in an attempt to understand the processes by which famine occurred at this time. According to the logic of Ester Boserup, the failure to improve techniques suggests either low population or unlimited land supply.[47] For the purposes of this enquiry, expansion was limited by the district's borders but cultivation had indeed been extended in the 1850s and 1860s onto poorer land. The state, too, believed that peasants should be looking to improve their farming methods, in preference to cultivating new land.[48] The only way to develop the rural economy and improve agricultural profits would be to invest in tools for more efficient and reliable production, particularly irrigation or more and better ploughs and bullocks. Rather than low population being the disincentive, they largely saw the problem as one of too many *ryots* who had little capital, bringing their instinctive Malthusianism to the fore.[49] While they claimed that low taxation was intended to encourage improvements and raise the 'standard of comfort' of the peasantry, it was seen that it had instead led to 'rapid ... appropriation of even the worst descriptions of waste'.[50] Thus the state appeared to be encouraging the wrong kind of development.

Neil Charlesworth has pointed out that, far from preventing such expansion or encouraging improved agricultural efficiency, it suited the government when more land was cultivated. Not only did it increase land revenue returns, it could be recorded as a sign of increased prosperity, in response to high prices and improvements in transport.[51] The Bombay Government, in contradiction of its later view, declared in 1871 that 'An increased revenue from the land is the natural result and ordinary index of material progress.'[52] This provides a perfect example of colonial ambivalence towards the agricultural sector – criticising peasants' actions as negative but claiming the same phenomena proved their own governance to be progressive. While the state might be credited with making subsistence farming possible for more people under Sir George Wingate's initially moderate *ryotwari* land revenue settlement in the 1850s, this was a strange thing to interpret as prosperity. Moreover, it behoved them to read signs of declining cultivation in the 1870s as their responsibility in the same way, and especially to avoid increasing taxation. Among historians, sanguine views of increased cultivation in the 1860s have been taken by Michelle McAlpin and Peter Harnetty, but refuted by Sumit Guha, who argues that it resulted solely from population pressure coinciding with relatively low taxation on poorer land. He concludes that the expansion of agriculture was at only half the rate of the growth of population in the Deccan and that the stretching of land, and also cattle, resources was a structural weakness of Deccan agriculture.[53] Increased cultivation was not matched by significantly greater output – indeed it explained poor average yields per acre in Ahmednagar, which is shown later – and drier land was farmed by new, poorer peasants, rather than existing holders. The central notion that *ryots* failed to develop profitable agriculture because they were lazy also required officers to ignore rural economic structures of which they were well aware. When Assistant Collector W. W. Loch suggested that low taxation had removed the need for efficient farming,[54] Revenue Commissioner William Havelock demolished his logic by pointing out that 'if the low rates have increased ... indolence, then it is an equally fair conclusion that the payments for debts to the Saokar over and above the assessment to Government is the antidote to that indolence'.[55]

Even if it had been possible for some peasants to accumulate capital, it was recognised that this was not possible for all, and would reduce the number of smallholders. The inconsistency of this goal with that of encouraging cultivation of utterly unprofitable land was papered over by differentiating peasants, not by the quality of their soil or other resources, but by their imagined attitudes towards opportunities for gain. On the verge of the famine crisis, in response to increasing resignations, the Government of Bombay intriguingly argued that polarisation due to its policies had increased the chances of famines for those unprepared to better themselves.

> Where people are content to live as long as the harvests are bountiful and then to die like sheep – a sense of individual distress is lost in what appears to be the overwhelming and destructive fury of fate; but it has been the object of the political economy of civilized life to recognise the laws which work in the natural world, and to anticipate

calamities by gradually removing the cause. A population steadily struggling away from poverty must leave many behind, and those who are thus at the base of civilized society, either because they cannot or will not work, suffer more acutely than if all society were content with a low standard.[56]

In the context of analysing the famine process, this amounted to a forceful argument for leaving things as they were. That the government did not see it thus exemplified the extent to which they failed to understand peasant realities or to base their agrarian policies around the need to avoid crises, as peasants did. Not only did the Bombay Government dismiss the significance of individual suffering, but they also insisted that the poorest were isolated by others' success, an argument which had little economic basis in Ahmednagar.

It may have been that increasing poverty was indeed simultaneous with a degree of peasant capitalisation in some districts. In Ahmednagar, however, the masses were being pushed down without the emergence of a noticeable rich peasant elite able to compensate by improving some land. Successive collectors reported that the majority of *ryots* were in worse condition than they had seen elsewhere.[57] The British may have been ambivalent on whether increased, undercapitalised, farming was an advantage. For Ahmednagar peasants, it was the norm that determined their modes of production and made it impossible to think of improving their use of land. Crop selection strategies such as rotation were only practical involving more expensive grains like wheat and *gram* (pulse), which would not grow on the driest soils.[58] The main foodgrains were the cheap millet *bajri* (*holcus spicatus*) and the sorghum *jowar* (*sorghum vulgare*). Many assumptions that existing methods were inferior were also based on lack of knowledge of local conditions. For example, *ryots* argued that manuring crops would backfire when rainfall was scanty, causing them to wither.[59] It was not worth gambling on maximising good yields at the expense of bad ones. Similarly, they sowed *jowar* and *bajri* thickly to ensure some return, though this reduced the grain yield per stem if they all grew.[60] Far from passively accepting their fate, cultivators constantly farmed in order to minimise the effects of low rainfall, at the expense of the chance to improve their profits. The British blamed them for their lack of dynamism, because they wanted them to be yeomen, but also because they associated poor conservative peasants with famine. This perception rested upon a false dichotomy. It was precisely the tendency not to speculate to accumulate which protected poorer *ryots* from the risk of famine – and the undercutting of this strategy, in part by the state, is thus of critical importance to the study of a famine period. This will be considered further in Chapters 2–4.

Livestock

There were more bullocks or cows per person in Ahmednagar than Poona, but not a high proportion per acre. In regions where they were more highly concentrated, like Karjat *taluka*, low human populations or large uncultivable tracts meant that they did not aid cultivation so much as compensate for its absence with dairy produce or by pulling carts.[61] Their importance to cultivation was considerable. There were only 64,855 ploughs in the district in 1876, but 23,204 needed two bullocks to pull them, and 41,651 four,[62] thus requiring half the available cattle, unless they were shared. It was, therefore, no surprise that the need to purchase (or replace) cattle was among the most common reasons given for *ryots* first getting into debt. Unlike land ownership and cultivation, livestock records do not show a decline in the early 1870s. They were particularly unreliable, however, being taken by village officers whom Boswell described as 'not over intelligent', adding, 'At present all such returns can only be looked upon in the light of rough estimates.'[63] When numbers apparently leapt in 1873–74, he insisted that the increase was 'wholly fallacious', reflecting attempts – including punishment of errant *kulkarnis* – to improve their accuracy.[64] The continuation of such efforts probably exaggerated the impression of post-famine recovery, as well as disguising earlier decline. The given figures nonetheless show how badly cattle numbers were affected by the famine crisis, as can be seen in Table 1.5.

During the famine, in addition to cattle starvation, Collector Stewart reported that many beasts were sold 'to butchers and merchants at minimal prices', in addition to extensive emigration of peasants specifically in order to keep their bullocks alive, 'from which they in all probability never returned'.[65] Woodburn also reported cultivators' claims that many good cattle, which did remain, had died after eating infected, withered *jowar*. Demonstrating a typical British attitude towards the credibility of the people under his charge, he added, 'I do not know whether this is true, but I have seen fine bullocks lying dead whose death was attributed to this cause, and have also been shewn the worm in the heart of the jowari stalks.'[66] Stewart warned that 'Scarcity of bullocks for ploughing will for some time to come weigh upon the agriculturists and paralyse their energy'[67] and this proved to be the case, with the animals remaining often being those considered too weak to take elsewhere.

Table 1.5 Livestock totals, Ahmednagar district, 1871–81

Year	Cattle	Horses/ponies	Sheep/goats	Carts
1871–72	497,469	18,193	364,542	22,015
1872–73	495,482	18,956	362,450	22,832
1873–74	511,400	30,668	400,041	23,167
1874–75	513,127	31,204	411,965	23,221
1875–76	579,115	32,337	462,050	25,053
1876–77	376,342	22,890	346,385	21,447
1877–78	381,385	22,868	388,438	22,264
1878–79	405,784	24,254	392,450	21,360
1879–80	434,722	24,940	415,897	21,802
1880–81	451,999	25,517	417,197	22,294

Sources: Boswell to Havelock, No. 1645, 20 July 1874, p. 116; Boswell to Havelock, No. 1952, 20 July 1876, p. 226; King to Robertson, No. 4584, 22–5 July 1881, p. 49; Elphinston to Robertson, No. 5730, 20 July 1882, p. 60.

However, in the context of imminent land revenue rises and after the expensive famine, the Government of Bombay was not keen to hear of continuing agricultural problems. For the next four years collectors fell back on the same unreliable village stock records as evidence of natural recovery, or even increased prosperity.[68] Even accepting this, there were fewer cattle in 1881 than there had been a decade earlier, when their scarcity had been noted. And in that year, there were outbreaks of both rinderpest and anthrax in parts of the district, killing only around 200 cattle, but affecting more.[69]

Levels of production

Levels of production in different *talukas* and years were recorded in two ways. While crops were growing, collectors, their assistants and *mamlatdars* produced estimates on the *annewari* system (for estimating relative crop yield), grading foodgrain output each season by *taluka*, from a perfect 16 *annas* downwards. Deccan Riots Commissioner C. W. Carpenter argued that this was unable to take full account of different levels of soil productivity, having calculated that good Deccan soils could yield as much as 26 times more than the worst.[70] Nonetheless, individual assessment presented a more realistic picture of local conditions than rainfall returns, and permitted a regulated method for consideration of claims for land revenue suspension or remission. Once harvests were in, *jamabandi* revenue reports gave detailed figures for acreages and yields. These are given for key crops in Table 1.6.

Table 1.6 Extent of cultivation (acres) and production (*maunds*) of key crops, Ahmednagar district, 1871–83

Year	Jowar	Bajri	Wheat	Gram	Cotton	Waste
1871–72*						
Area	1,123,862	275,411	109,947	59,703	411	593,756
1872–73*						
Area	703,462	1,028,609	123,325	78,034	9,616	282,805
1873–74*						
Area	839,667	781,734	183,384	75,268	20,055	274,657
1874–75						
Area	648,176	923,186	142,046	87,049	48,063	314,814
Yield	533,104	781,169	153,905	97,260	60,327	—
1875–76						
Area	675,002	874,815	152,978	85,488	34,989	148,112
Yield	598,544	672,681	190,732	95,382	47,572	—
1876–77						
Area	422,515	796,625	49,997	20,886	13,297	162,897
Yield	153,889	265,228	40,604	10,680	12,360	—
1877–78						
Area	831,190	721,526	118,490	39,844	16,528	469,137
Yield	364,992	406,971	46,630	17,091	18,250	—
1878–79						
Area	544,487	1,034,894	92,220	44,997	20,421	436,258[†]
Yield	341,790	1,532,400	81,336	39,495	19,229	—
1879–80						
Area	650,689	806,148	104,597	52,652	20,550	407,670[†]
Yield	449,999	533,845	113,427	61,940	21,478	—
1880–81						
Area	917,958	528,713	171,961	90,425	11,055	—
Yield	893,059	520,097	177,801	84,222	18,403	—
1881–82						
Area	679,879	783,150	151,026	64,470	32,231	319,901
Yield	603,520	607,345	114,613	63,225	177,591	—
1882–83						
Area	599,643	889,162	135,335	58,154	88,020	—
Yield	692,759	583,436	230,672	85,826	174,623	—

* Yield statistics not given.
[†] Figure for total cultivable land minus all crop totals, thus including roughly 10,000–20,000 acres cropped twice.

Sources: Data from annual *jamabandi* reports as follows: Havelock to GOB, No. 4799, 9 December 1873, MSA, GOB, RD, Vol. 11 of 1874, No. 62; Oliphant to GOB, No. 4619, 21 December 1874, MSA, GOB, RD, Vol. 12 of 1875, No. 543, pp. 320–7; Oliphant to GOB, No. 539, 24 February 1876, pp. 110–11; Robertson to GOB, No. 3654, 4 August 1877, pp. 88–107; Robertson to GOB, No. R/626, 16 February 1878, pp. 11, 68–83; Robertson to GOB, No. R/336, 8 December 1879, pp. 448–59; Robertson to GOB, No. R/624, 25 February 1880, pp. 280–92; Robertson to GOB, No. R/676, 17 February 1881, pp. 338–49; Robertson to GOB, No. R/817A, 25 February 1882, pp. 78–83; Robertson to GOB, No. R/1735, 10 April 1883, pp. 72–7; Robertson to GOB, 14 May 1884, pp. 72–7; *Ahmadnagar Gazetteer*, p. 245.

The degree of variation between years is considerable, especially the relative figures for *jowar* and *bajri*, confirming that cultivators were flexible in crop selection according to conditions. For example, there was a remarkable discrepancy between the areas of cultivation of *jowar* and *bajri* in 1871–72 and 1872–73. *Jowar* was generally a winter crop in Ahmednagar, sown for the later, *rabi* harvest. Thus, in 1872–73, as in most years, fewer acres of *jowar* were sown, reflecting a respectable first, *kharif*, harvest, mainly of *bajri*. By contrast, in 1871–72, when the *kharif* rainfall was very poor, four times fewer acres of *bajri* were reaped than of *jowar*. As these figures are for acreages, not yields, they reveal that cultivators had switched in mid-season to an unusual emphasis on the *rabi* crop, in many cases by uprooting the failing *bajri* before the harvest, to start again. This sometimes meant that rain late in the *kharif* season which could have revived the crop served no productive purpose.[71] Some land was also better suited to particular crops, producing diminishing returns when alternatives were sown.[72] As a result, yields vary rather more than acres sown, particularly in the dramatic decline of 1876–78. Nonetheless, this suggests, importantly, that *ryots* continued to attempt to cultivate as far as possible, even in the most adverse conditions, and when fewer seedgrains were available.

The early 1880s figures give the impression of slight recovery, with increased yields per acre of *jowar* (but not *bajri*) and increased cultivation of more marketable crops suggesting greater resources to spend on seeds. Greater cotton production may also have been in response to the construction of the Dhond–Manmad railway. It is not possible to tell how many *ryots* benefited from this, given that district level figures do not show up the extensive crop failures in northern *talukas* at this time. It may have been that *Marwaris* who called in mortgage bonds in response

Table 1.7 Average yields (*maunds* and *seers*) per acre cultivated, Southern Division, Bombay Presidency, 1874–5

District	Jowar	Bajri	Wheat	Gram	Cotton
Ahmednagar	0–33.5	0–35	0–43.5	0–43.5	0–51.5
Poona	1–25	1–22	2–0	2–46	1–38
Satara	1–38	0–64	1–49	—	2–2
Sholapur	0–77	0–52	2–13	1–21	1–24
Belgaum	2–55	2–18	1–76	—	0–40
Dharwar	2–66	2–8	2–29	2–4	—

Source: Oliphant to GOB, No. 539, 24 February 1876, pp. 128–9.

to the Deccan Agriculturists Relief Act turned the land to cash crops, using hired labour. Moreover, average yields per acre needed to improve in Ahmednagar, as comparison with other districts in the 1874–75 *jamabandi* report made unimpressive reading, as shown in Table 1.7.

Rainfall and drought

Low levels of production per acre confirm that natural conditions were poor for cultivation in Ahmednagar. The particular problems of the district were recognised by the Government of Bombay, which recorded that

> The western portion of the Dekkan, including the districts of Nasik, Puna, and Satara, is hilly; the valleys rich and highly cultivated; and the country diversified and beautiful. Further to the east the country is more level, the soil of a more arid description, and much less productive, owing to the diminished rainfall.[73]

Alongside Sholapur to the south, Ahmednagar district, in which only the isolated tribal *taluka* of Akola in the northwest was hilly and forested, was thus excepted from – perhaps counterpointed to – attempts to shed positive light on the Deccan plateau as a whole, in which rainfall failure was not unusual. Three years before the famine crisis, it was noted that 'almost all branches of the administration are in some degree affected by the season, and in the explanation of events a constant reference is made thereto'.[74] Ahmednagar had always been prone to serious rain failures and famines, with 12 recorded in its history, seven of them in the nineteenth century, and five in 50 years since the onset of British rule: in 1824, 1832–33, 1845–46, 1862 and 1876–78.[75]

The Famine Commission's view that Indian famines were exclusively caused by drought was reinforced by an appendix reporting a lecture by its chairman, Lieutenant-General Richard Strachey, to the Royal Institution of Great Britain in 1877, which considered W. W. Hunter's suggestion that famine could be predicted by observing sunspots. Strachey offered the 'Indian' philosophical observation that life is based on struggle between forces of preservation and destruction, with climate a particularly potent such force.[76] His lecture is praised as 'informative' by B. M. Bhatia, who himself asserts that 'The immediate cause of famine in [India] is almost invariably drought or unseasonal rains.'[77] Rainfall failure can be accepted to have been the trigger of most Indian famine crises but not their sole cause. Moreover, the problem remains of how drought was measured in contrast to perceived norms.

Detailed statistics were kept during the famine, but their compilation underlines British assumptions. In a preliminary memorandum summarising the Government of Bombay's intended strategy, collectors were instructed to submit weekly rainfall data, along with observations on the availability of water, the state of crops and cattle, prices and relief measures. The condition of people, and mortality, were conspicuously omitted from the list.[78] Throughout the famine, the government's weekly statements laid emphasis on rainfall returns, matched in significance only by expenditure figures. While this served an administrative purpose, such data is of little comparative use when rainfall was more crudely recorded at other times, and its more local impacts overlooked. Yet the frequency of scanty rainfall in Ahmednagar was well known, including a common estimate that severe scarcity would strike the tract every 11 years, and famine every 50.[79] In 1873, Havelock suggested that that year's good season should be regarded as unusual, adding, 'I know by long experience that it is a very rare thing for the rain to fall copiously and seasonably.'[80] Annual rainfall totals, broken down by *talukas*, are given in Table 1.8.

Although this reveals how capricious rainfall patterns could be in the district, it provides only a rough insight into the considerable spatial

Table 1.8 Range of average annual rainfall measurements (inches), Ahmednagar district, 1869–82

Year	Highest *taluka*	Lowest *taluka*	Average
1869	Ahmednagar: 47	Newasa: 22	31
1870	Parner/Akola: 19	Newasa/Sangamner: 5	10
1871	Parner: 25	Sangamner: 6	13
1872	Karjat: 32	Sangamner: 15	22
1873	Ahmednagar: 33	Karjat: 15	21
1874	Sheogaon: 36	Rahuri: 22	28
1875	Shrigonda: 39	Karjat: 15	25
1876	Sheogaon: 21	Karjat/Sangamner: 7	10
1877	Karjat: 30	Akola: 12	19
1878	Jamkhed: 40	Shrigonda/Kopargaon: 24	30
1879	Newasa: 37	Shrigonda: 25	25
1880	Jamkhed: 27	Kopargaon: 12	18
1881	Jamkhed: 24	Kopargaon: 8	18
1882	Karjat/Jamkhed: 32	Kopargaon: 17	25
1860–82 average	Rahuri/Jamkhed: 26	Sangamner: 17	21

Source: *Ahmadnagar Gazetteer*, p. 13.

variations of rainfall within the district, and none into temporal ones within each year, which could be critical. Moreover, while it shows 1876 to have had very poor rainfall, it was no worse than 1870, which was followed by worse levels in 1871 than those experienced in 1877. The worst individual *taluka* figures were also in 1870–71. While relief was provided and mortality reported that year in Ahmednagar and Nasik districts, the local nature of the scarcity prevented famine from being declared – or indeed its possibility from being acknowledged. Similarly in 1881, Kopargaon had almost as little rain as the worst *talukas* in the famine, but local relief the following year attracted little notice. Thus even this short run of average figures suggests that the monsoon failure of 1876–77 could scarcely be labelled as 'unusual drought', to fit in with the Famine Commission's subsequent definition of famine. It was well below average, and perhaps below a critical point, but not by so much as to be rare or especially extreme, given a high standard deviation in annual rainfall levels over time. The *Ahmadnagar Gazetteer* confirmed that scarcity was normal in the district, reporting 'In the plains the early rains are often scanty and the late rains capricious, so that droughts ... appear to form the rule and a good year the exception.'[81]

Though the significance of the uncontrollable rains was over-emphasised after the event, opinions at the time of the famine were more restrained. Jacomb described the 1876–77 season merely as 'more or less quite unfavourable',[82] while the Government downplayed Ahmednagar's difficulties: 'Though the rain-fall was very deficient, yet this district cannot be classed in quite the same category as some of the other Deccan Collectorates for in five or six Talukas there was a fair yield of crops which *pro tanto* averted distress.'[83] Relative analyses thus served Ahmednagar poorly in different ways. In 1870–71, there was no crisis because rain failure extended to only one other district and in 1876–77, there was less cause for concern because it was worse elsewhere. But this logic was not applied internally. Those *talukas* which did have very little rain in 1876 were discounted by aggregation. Playing down the significance of local droughts in Ahmednagar in both 1876 and at other times shows poor understanding of the way in which rain failure impacts upon individuals.

The Governor of Bombay, Richard Temple, argued, further, against the view that a single crop failure 'plunges the people into distress, or draws them to the verge of pauperism', adding that low rainfall for a second year in 1877 'prolonged in an aggravated form the consequences of the first failure'.[84] This supported an emerging opinion that famine resulted only from two years of rain scarcity – a definition which conveniently left the blame on the monsoon, but also categorised famine as rare and justified late intervention. As has been seen, however, consecutive rain failures in 1870–71 were not called famine. Moreover, there had been five consecutive years of poor local rainfall in the 1860s, but crisis was averted by high agricultural prices caused by the American civil war.[85] This is consistent with Sen's argument that food availability decline is neither necessary nor sufficient to cause famine.[86] Colonial understandings of what constituted famine rested arbitrarily on the extent and depth of suffering required to convince outsiders that it was occurring, while serving to deny chronic vulnerability. Yet official reports confirmed considerable distress in Ahmednagar as early as October 1876, with widespread migration and over 20,000 people on relief, which was to double by early December.[87]

Even if rain failure is accepted to be a trigger of famine and chronic vulnerability is discounted, the problem needs to be examined locally. The argument that it should be geographically widespread and long-lasting to count as famine was at odds with the nature of the rainfall problem, which was precisely that it varied considerably over short distances and periods, as well as affecting the land in disparate ways. For example, the northwestern *talukas* were adjudged by Loch to have had a good season in 1874–75, despite the facts that the late rain was insufficient to allow any *jowar* yields on lighter soils, and that Kopargaon had had 27 per cent less rain than it had averaged in the previous five years.[88] Even aggregated figures for *talukas* could mislead. Collector John Elphinston explained his reports of impoverishment in Parner *taluka* in 1882, when the rainfall figure was well above average, by pointing out that the statistic was 'taken in the pluviometer at Parner [town], so it is possible that villages at a great distance N.E. of the mamlatdar[']s headquarters may have suffered from a want of rain for the rabi crop'.[89] The previous year, King described the effects of uneven rainfall within single villages: 'crops of the same kind but in all stages of development might be noticed almost side by side and the stage of maturity was reached with ... little uniformity'.[90]

Perhaps more importantly, annual rainfall figures also concealed the temporal distribution of rainfall, which was crucial to the success of the harvest. Again this problem was well known locally, but could be lost when considered in less detail at higher levels. The famine provided a striking example of this in Shrigonda *taluka*, where the total rainfall for 1876–77 was 15.87 inches, 73 per cent of the recent average, like Kopargaon two years earlier, and a deviation from the norm too common in a single *taluka* to warrant comment in many years. Yet the

Table 1.9 Rainfall figures as a percentage of monthly average, Eastern Deccan, 1876–77

Year	June	July	August	September	October	Whole year
1876	55	111	35	31	5	47
1877	149	38	95	268	183	142

Source: Famine Commission Report (FCR) (1880), 'Bombay Information and Evidence', chapter 1, question 1, answer by F. Chambers, Meteorological Officer.

extent of crop failure and distress in Shrigonda was reported to be the highest in the district. Closer examination reveals that 11.84 inches had fallen in June alone, 2.98 in July and only 1.05 in the next eight months, causing all crops to die.[91] Similar discrepancies throughout the eastern Deccan were noted by Meteorological Officer F. Chambers, who calculated rainfall as a percentage of the average figure for each month, as shown in Table 1.9. Aggregating a large region, rain was thus above average at the height of the 1876 monsoon, failing most dramatically only later. Conversely, the figures appear good for 1877, but the rain failed at a critical time after the *kharif* planting, discouraging many cultivators from risking their last seedgrains when conditions were good for the *rabi* sowing, thus further reducing the correlation between total rainfall and output.

Such geographical and temporal rainfall variations were not only recorded during the crisis but frequently, particularly in the years after it, when peasants' lack of resources increased their aversion to gambling on cultivation when the monsoon had started badly. In southern *talukas* in 1878, for example, when recovery was assumed by the government to be under way, respectable overall rainfall figures were so unevenly distributed that some villages grew barely any crops at all.[92] Cultivation was also limited by the lack of seedgrains and bullocks for ploughing, with many fields left grassed over in the hope of recovering lost capital in the fodder market. By contrast, in the same year in the north of the district, rain failure in July, August and November killed most of both *kharif* and *rabi* crops after widespread planting.[93] The constant variability of the rains confused British overseers by affecting yields and farmers' choices in different ways – sometimes favouring flexibility and at other times caution. In 1880, Assistant Collector Richard Candy retrospectively noted in his *talukas* that the best rains had been earlier than anticipated, in May, while the later monsoon had been inadequate, and the *rabi* harvest damaged by frost. The *ryots*, he concluded, had been foolish not to take advantage of their 'splendid' fortune at the start.[94] That they had not known how the rest of the season would turn out did not prevent them from being blamed. Nor was insufficient rain the only difficulty with which agricultural producers had to contend. In several years, excessive precipitation caused damage to young crops or delayed sowing. It could also lead to other problems, including weeds, locusts, other destructive insects called *kharpadras*, and plagues of rats.[95] In 1880, the government paid out rewards totalling 16,573 rupees for killing rats, in a rare measure of direct support for agricultural production. Collector King claimed that this resulted in the death of 1,767,414 rodents throughout the district,[96] but Candy undermined the bizarre precision of this statistic by remarking that 'the Kunbis could not be induced to join in the work of destruction' and that rats had 'mysteriously disappeared' equally in areas where large and small rewards had been claimed.[97] Dry spells, too, encouraged crop diseases on several occasions, particularly a *jowar* blight, *khyri*.[98] In addition, frosts, such as that reported in 1880, were commonly harmful, and a cyclone passed through Jamkhed *taluka* in November of the same year, bringing over five inches of rain in a single day.[99]

It is interesting to note that, although none were new, many of these problems arose in the years after the end of famine relief, sometimes coinciding in single *talukas* and years. This may have related to the reduced extent of cultivation due to the weakness of the population or continuing bad seasons, or alternatively reflected greater attention to farming problems in administration reports. It does, however, underline the fact that, while agricultural production could be beset by rain failure and other natural difficulties, they were not much worse in the famine years than at other times, and in some *talukas* less so. Chronic variability in the conditions for food production in Ahmednagar meant that the 1876–78 famine cannot be explained wholly even by greater degrees of variation.

The impact of the famine crisis

Peasants were struggling to survive before the famine, although their case was not seen as pressing by the higher echelons of the state. Their chronic difficulties worsened during the crisis itself. Jacomb reported in 1877 that 'It is feared that for a few years to come at least the District will probably show a marked decrease in cultivation.'[100] Occupancies also declined further as resources were stretched, in addition to the recognition of famine departures. Woodburn argued in 1878 that less than half the area counted as cultivated in Shrigonda and Karjat *talukas* had been planted in reality.[101] The Government of Bombay was not keen to hear such negative evidence, seizing instead on any hint of optimism as they surveyed the reports from below. Some local officers were keen to confirm the colonial fallacy that the famine had been a sudden, unexpected event, out of kilter with normal patterns of existence, by talking up the *ryots*' prospects in its aftermath. Such optimism created tension between its exponents and others who argued that a realistic appraisal of local conditions rather necessitated government measures to prevent near famine conditions from becoming chronic. This process highlighted the antagonistic way in which opposing perspectives were often played out at different levels of the state hierarchy. In 1877, for example, Hamilton reported that conditions were 'materially favourable' in his *talukas* in the north and west of the district, and the effects of the famine were over. The 'very considerable' revenue increases already levied in Kopargaon *taluka* were warranted, he suggested, by the underlying fertility of the soil, which made profits potentially large 'if rainfall is good'. This was reflected by the take-up of land abandoned during the famine in Kopargaon, and although this had not yet happened in the neighbouring *taluka* of Sangamner, he expected it to soon.[102] While Kopargaon does have the best soil in the district, and was less affected by famine in 1876–77 than southern *talukas*, Jacomb implied that the point had been wilfully missed:

> Mr. Hamilton apparently takes a sanguine view of the general condition and prospects of these Talukas but considering the nature and extent of the failure of crops last year and the consequent scarcity the Acting Collector apprehends that there will be a falling off in cultivation if as has been the case in the other affected Talukas there has been a loss in the No. of cattle.[103]

It was inappropriate, furthermore, that Hamilton should have focused attention on Kopargaon rather than Parner, another *taluka* under his charge, which had 'deteriorated in its general condition and prospects'. Jacomb had visited Parner himself, but not Hamilton's other *talukas* during the last year, and admitted that he could not, therefore, comment on their 'actual state ... from personal experience'.

Close observation was thus valued over assumptions or generalisations, but this was counted against Jacomb by Robertson, who similarly ignored Parner in concluding that, 'Being on the spot and having seen [his *talukas*] through all the worst times, [Hamilton] has had a better opportunity of forming an opinion on the subject than the Collector.'[104] This suited the Government of Bombay, which further overlooked Hamilton's own equivocation regarding Sangamner (and a pessimistic view of trade in Akola), declaring that it would 'rather incline to the more favourable view taken by the Revenue Commissioner in his remarks on the report of the 2nd Assistant Collector.'[105] Therefore, the desire to believe any record of recovery overrode, for the provincial government, both local observation and ambivalence. Such colonial discussions were too often predicated upon the prognosis of improved natural conditions. Hamilton's optimistic attitude relied on the coda that rainfall had to be good, which it did not turn out to be in Kopargaon for several years after 1877, yet climate was not mentioned once in any of the subsequent commentaries. At a still higher level, though, the famine crisis had the converse effect of shocking the distant Government of India and Secretary of State out of their complacency regarding general conditions in arid zones like the Deccan. Both put considerable pressure on the Bombay Government to reduce their land revenue demands in Ahmednagar and to investigate the conditions of the district more seriously. Lord Cranbrook commented on the same year's Presidency revenue report, for example, that, 'Although it appears ... that the cultivable and assessed area was larger ... much of the so-called cultivated land did not come under the plough.'[106] Hamilton's data on land taken up in Kopargaon contained precisely this flaw, including land re-categorised from cultivated waste to cultivable, but not excluding occupied land left fallow.

Similar optimism to Hamilton's was common in reports of recovery after the famine. Rainfall and cultivation statistics were taken as positive because they were better than the crisis years, even if mid-term trends were still poor. For example, in 1879, before famine resignations had even been recognised in the statistical record, Candy and King were celebrating modest increases in cultivation.[107] The following year, when the records dropped dramatically, their significance was denied, as has been seen, and replaced by unwarranted optimism, without evidence, that large areas of waste would soon be taken up by well-to-do cultivators.[108] The first aggregate increase in cultivation after the famine eventually came in 1882, but even this was assisted by re-categorisation of land under the revenue survey. The desire to paint a picture of recovery was still greater than the evidence would allow, with Assistant Collector N. G. Deshpande, for example, declaring that prospects were favourable due to increased cultivation in all his four *talukas*, when his data showed slight decreases in Parner and Jamkhed.[109]

Robertson sounded a note of caution, suggesting that, while the figures showed *ryots'* willingness to take up land in a decent season, it was, again, 'often more than they can properly cultivate'.[110] Thus new take-ups were inferred to be by poor rather than rich peasants, and recovery amounted to no more than a return to the conditions in which famine had struck. Other officers supported the view that the agricultural situation remained vulnerable in Ahmednagar long after the famine, even as the Government of Bombay persistently concluded that it was improving. With state famine relief having failed to support agricultural production, devastated *ryots* were left to pull themselves up by their bootstraps afterwards in extreme circumstances. Woodburn described how those who had avoided migration or relief works had 'clung to their hearths and homes, and earned a scanty subsistence by cutting and selling firewood. The destruction of trees has in consequence been enormous.' For such elementary physical reasons, he argued, it would take 'some years to recover the condition of twelve months ago'.[111]

With harvests continuing to be poor, due to successive scanty monsoons and reduced resources, Hamilton and Candy had become pessimistic by 1880. Hamilton noted that people had become disheartened, relying only on the hope of a good season.[112] This was disputed by King, who held that people, especially in tribal areas of Akola, had been 'described as stricken with almost inconceivable poverty' before the famine too.[113] Not only was this scant consolation, it was disingenuous when Hamilton had been referring to almost half the district, including the supposedly prosperous *talukas* of Kopargaon, Rahuri and Sangamner. In non-tribal Shrigonda, meanwhile, Candy reported that 'you may walk for many miles and not see a cultivated field or a human being. In the villages many of the houses are in ruins and untenanted. In fact whole families disappeared during the famine and never returned.'[114] By 1881, Kopargaon and Shrigonda were both being described as 'desert', and King anticipated renewed distress.[115] Optimism could only be derived now from differentiating *talukas*, with Deora arguing that whereas southern *talukas* like Shrigonda were chronically poor, Sheogaon and Newasa to the east were 'rich and productive', because their 'cultivating class consists of well-to-do ryots'.[116] Yet in 1874, Boswell had spoken gloomily of the depression of the cultivating classes in the same *talukas*.[117] Kopargaon, on the other hand, had been regarded as a fertile and prosperous area when its revenue was the first to be revised in 1876 and 1877. It was predicted to benefit from the construction of the Dhond–Manmad railway – which passed some distance from Sheogaon – through its midst. Such variation in fortunes thus merely serves to emphasise the uncertainty that prevailed throughout Ahmednagar; no part could ever be regarded as safe from famine. In 1882, Elphinston, the new Collector, was obliged to open relief works in Kopargaon, and offered a still gloomier view of the general condition of the peasantry. In the face of persistently low yields and prices, and rising land revenue, he wrote,

> small farmers are in emergent need of money. They have lost oxen by disease, or sold some (or even all) of them to pay the Government assessment and for food for themselves and fodder for their remaining oxen, so they need oxen. They need seed for sowing their fields &c &c.[118]

Apparently frustrated by the indifference of his superiors to this, he was moved to ask, 'How is a poor cultivator in such a state of things to recuperate his resources and rebound to the position he held before this long famine ...?'[119] At least one British officer, therefore, conceived of famine as a chronic process lasting for years, not just one or two seasons of unusually low rainfall.

Local officers were not always willing to report peasant poverty, and when, like Elphinston, they did so too strongly, they earned criticism from above.[120] Yet there is enough in colonial records to conclude that Ahmednagar *ryots* were in chronic difficulty throughout the period 1870–84, and not just during the declared famine. Soil and rainfall were inadequate to sustain a heavily undercapitalised and indebted peasantry. It has been suggested, moreover, that higher levels of government were not concerned with poor cultivators. This chapter concludes by investigating state agendas for peasant cultivation in Ahmednagar.

Government agendas for rural development

After the famine, the Bombay Government claimed that in the central Deccan, 'the impoverished condition of the cultivating classes has been for some time anxiously and attentively watched by Government with the view of introducing some means for its amelioration'.[121] While this was reflected in the Deccan Agriculturists' Relief Act, it was an attitude of pity in peculiar circumstances, rather than any statement of intent to uplift the peasantry. Government reactions to local reports of difficulty remained as dismissive and overly optimistic after the famine as before, when they had held that leaving many to suffer was a fair price for progress. Indeed, if collectors' and assistants' Annual Administration

Reports are read in sequence throughout this period, the emphasis under the heading 'General Condition of the People' switches from poverty to the state of harvests to prospects, which were said in 1878, for example, to be good solely because of the new railway.[122] Only the renewed crisis of 1881, and Elphinston's personality, returned the focus to the struggling people. The lack of a coherent understanding of the problems faced by *ryots* was exacerbated by the rapid turnover of local officers, such that in the entire period, only Boswell and King submitted more than one annual collector's Administration Report. Assistants were moved still more frequently, prompting Boswell to complain at length that 'These constant changes are a great hindrance to work and a still greater evil is that they leave no one in a District, who really knows the country or its people.'[123] Though staff turnover was normal within the Revenue Department, he implied that Ahmednagar and its people were not regarded as a high priority in Bombay, where it was assumed that anyone demonstrating competence deserved a more comfortable post. Boswell complained particularly at the removal of Loch, who 'was actively engaged on enquiries which would throw light on [the district's] true condition and has taken a deep interest in the matter'.[124] In addition, he argued that the decision to dispense with native *dufterdars* (clerks) who had provided continuity in 'sifting and weighing properly the often conflicting and inaccurate reports received from the various Mamledars' was a false economy if the government required meaningful information.[125] No statement of concern from the provincial government changed this situation, with Stewart confessing in his 1878 report that he had had to ask his colleagues in order to get a picture of the condition of the district,[126] and King complaining in 1881 that for ten months of the year he had had only one assistant collector because of shortages and leave, when only Khandesh district had more work to get through in the entire Presidency.[127]

More important than levels of staff or attention, agriculture in Ahmednagar suffered from a lack of investment. When the Famine Commission enquired about help given to improve tools, seed strains or methods, Robertson told them, 'Nothing has been done by private individuals and next to nothing by Government. On the contrary, this most important and pressing duty has been most grossly neglected by Government.'[128] State rural investment went into occasional large scale transport or irrigation projects, not to meet smallholders' needs. Only small *takavi* loans were offered for seeds, bullocks or wells, as examined in Chapter 3. When the state experimented to improve farming itself, its aim was to introduce profitable crops, without any success in

Ahmednagar. Attempts to grow sunflowers from seeds imported from Dharwar were described as not even encouraging, Boswell observing, 'This is just what might be expected in a climate so dry as compared with that of the Southern Maratha Country where the plant is said to flourish.'[129] Mahogany, carob and tamarind were also tested, on a scale described by Robertson as 'too trifling to require more than a passing notice'.[130] The only other government sponsored assistance to cultivators in the district was the keeping of a handful of stallions for breeding, by the Superintendent of Police.

In the early 1870s, in response to a Government of India memorandum circulated by A. O. Hume, collectors were asked to consider the value of model farms to conduct similar experiments and educate *ryots* in improved techniques. W. D'Oyly in Ahmednagar replied that there was an ideal site at Farriah Bagh, though smaller than proposed by Hume at just 250 acres, which could be established for around 20,000 rupees.[131] The Government of Bombay was more circumspect, telling Hume that four or five farms in the whole presidency would suffice.[132] In the event the three that were created, in Khandesh, Dharwar and Sind, made huge losses from the start and were threatened with closure within two years if they could not cover their running costs.[133] Thus agricultural research and development was on an occasional rather than a continuous basis. The need to sell produce to survive discouraged model farms from experimenting too radically or testing measures to protect food security. The Dharwar farm was abandoned after three poor seasons. Robertson poured further scorn on this muddled thinking, arguing presciently for proper investment in an agricultural college, with small branches in every district, to research ways of helping farming in local conditions, based on indigenous knowledge. 'We want to teach the natives,' he declared, 'but have first to learn all they know, prove it, and try to improve upon it.'[134] Model farms on a scale that bore no relation to peasant farming, experimented with cotton strains and eucalyptus trees but not foodgrains, were nowhere near Ahmednagar and starved of funds, could never do such a thing.

Even if Robertson's plans had been followed, they may not have helped much. A small scale attempt was made, in 1880, to teach agricultural methods to high school children near Ahmednagar city, with scholarships provided for cultivators' sons and even a ploughing competition to generate interest. King revealed, however, that there were few *Kunbi* applicants, most of whom found pretexts to return home after a short time, encouraged by families who did not trust the government, 'being apt to think that they know quite enough about ploughing and harrowing already + that they might as well practise those accomplishments at home for themselves, as work like labourers on the farm at Nagar.'[135] As a result, soil science was taught only to regular schoolgoers, who had 'a great ineptitude for handling the plough or any of the practical work of the cultivators'.[136] This was symptomatic of the lack of educational provision of any kind for cultivators, while traders and moneylenders sent more of their sons to schools. Boswell complained in 1874 that not only were there a mere 129 primary schools covering the district's 1370 towns and villages, but also that four had recently been converted to senior schools. This was a wrong priority, he argued, 'while so little is done for the masses. These have the best claim on us both because they can least help themselves and because they pay most of our taxes. They are also the class whom Government profess to be most anxious to instruct.'[137]

Part of the reason why local officers – and even Revenue Commissioners Havelock and Robertson – so often found themselves contesting the agricultural policies of their government was that, as members of the Revenue Department, they had a remit to promote the cause of their tax payers, which was not shared by other government departments. They clashed particularly with members of the Survey Department over revenue increases. The latter department's control and influence over land measurements and categorisation, when key decisions were made, had a profoundly harmful effect on the government's attitude towards the peasantry. Its ultimate replacement by an Agricultural Department, which had to consider *ryots*' well-being, as recommended by the Famine Commission Report, may have had an ameliorating effect after the end of the period under study. There were different perspectives on matters concerning peasant agriculture too, between the Departments of Revenue, Forestry and Public Works.

The Conservator of the Northern Division told the Famine Commission:

> Rainfall, its capriciousness in the Deccan and in parts of the Presidency not immediately bordering the sea, is undoubtedly due to the destruction of woods and forests ... When the plains of the Deccan and Khandesh grew trees, then the fields were necessarily more fertile than they are now. The trees furnished the material for the formation of the vegetable mould from which crops derive their nourishment.[138]

On this Revenue Department officials were agreed, their frequent comments on the barrenness of Ahmednagar conflicting only with the

Survey Department's optimism. Too little was done to rectify this situation, however, before or even after the famine. The problem faced by the Forestry Department was again the requirement to pay its own way. Thus the costs of re-planting, and the considerable ones of re-marking the area under their jurisdiction, had to be recouped by selling timber and other forest produce.[139] The logic of this was challenged by Boswell, who argued in 1875 that it would be wise if felling were to be suspended in Ahmednagar for a few years and 'instead of showing a profit Government were to expend (were it possible) 10,000 Rupees a year in rearing new Forest. I know no district in the Presidency anything like so bare of wood as this is.'[140] Forestry income continued to exceed expenditure, however, prompting Boswell to attack the Forestry Department directly. In response, the District Forest Officer claimed that he only ever cut trees which were deteriorating, to which Boswell remarked 'I confess from such of his work as has come to my notice to being very sceptical as to his carrying out his own theory.'[141]

The famine crisis greatly upset Forestry balances. Not only were receipts reduced by lower demand for wood and the declaration of free forest grazing rights, expenditure was also increased by the employment of workers on relief to cut down prickly pear in forest areas.[142] This being a common device by which the Bombay Government reduced the apparent level of their famine spending by dispersing it under ordinary administrative heads, the deficit was overlooked. It was more surprising that a continuing shortfall was also smiled upon the next year, when all other departments had fallen back into line. The official view was that 'In districts like Ahmednagar Government can only look to the general good and to the progress of years for compensation of the excess of expenditure over receipts under the head of forests.'[143] This would have been a sound policy over several years – albeit not the most obviously helpful gesture for the stricken *ryots* – assuming that the extra expenditure was being used for reforestation. However, in the short term much of it was taken up by the need for extra work demarcating the Forest Department's boundaries, after large amounts of extra land – some of it abandoned by *ryots* – had been allocated to them, thus reducing the embarrassing levels of waste land in the district.[144] King pointed out that this work was preventing any new planting, and went as far as to argue, in a rare fair season, that, until the government could supply resources for that, the department should 'resign for cultivation under certain conditions, such tracts as have no natural growth of trees'.[145]

Ironically, the following year, King backed a suggestion by Candy that still more land might usefully be taken for forest reserve in Shrigonda and Karjat as so little was being cultivated or grazed.[146] He did add the pointed proviso, however, that if it was the Forestry Department should try to grow forests on it and not grasses, 'which any Patil could manage', as was then being done.[147] Robertson insisted that the full work of demarcation had to be completed before either re-cultivation or reforestation were permitted. Thus, on a rare occasion that the colonial primacy of economy was breached, it was only for the sake of their second mantra of bureaucratic efficiency. A Forestry Department too short-staffed to deliver adequately on either count was not going to revive the soil or rainfall levels in Ahmednagar, only reduce the cultivable area. Most of the trees planted in the district in this period were under a programme to provide shelter by roadsides. With saplings repeatedly wilting in the sun, this proved as unsuccessful as it was misdirected, the nadir coming when Stewart recorded 'several cases of destruction of existing road side trees which have been carried out in a very barefaced manner', presumably for firewood.[148]

Irrigation was an even more important area in which state assistance was needed if cultivation was to be improved in the district. There were difficulties in building large projects in the Deccan, as much land was prone to salination if over-watered and irrigation streams prone to rapid silting. Moreover, irrigated land required both more labour and manure, as well as incurring considerably higher land revenue.[149] Loch suggested that it was of no value unless it made cultivation of cash crops like sugar and plantains possible.[150] The *Ahmadnagar Gazetteer* reported that channel watering was very profitable but rare because of poverty, and irrigation use reflected this. State irrigation works were capable of supplying water to 41,510 acres of fertile land where rainfall was unreliable, but so few *ryots* were able to afford it that only 457 acres were irrigated in 1875.[151] Private wells cost from the equivalent of £10 to £500, depending on their depth and quality. Of 26,306 wells counted in Ahmednagar, only 1,718 were fully built with steps.[152] The district's rivers, Godaveri, Sina, Mula and Bhima, could all be dammed jointly by villages with *bandharas* (dams), but these required annual re-building and after the famine were reported to be widely ruined.[153] However, irrigation usage was far higher in seasons of drought, and extensive state irrigation schemes could potentially have saved the government considerable sums in lost revenue and famine relief costs, even if they had themselves been loss-making. Instead the 1870s saw declining investment in irrigation in Ahmednagar.

Two canals, the Ojhar and Lakh, supplied the district, as well as Bhatodi Lake, built by *Nizam* kings. The Bombay Public Works Department spent around 250,000 rupees each on the latter two in the early 1870s. The improvement work at Bhatodi was sufficient to mitigate the effects of the 1871–72 drought, but no work was done that year on the Lakh canal, although it was close to completion, with the result that only adjacent land could benefit.[154] Even when it was finished, it was only 21 miles long, offering water to 24,000 acres of poor, underpopulated land, so the take-up was minimal. The Bhatodi tank served a mere 14,000 acres, and was little in demand when the monsoon was full.[155] After spending large sums, the government were not pleased to receive such low returns on them and, in 1875, 'Under orders from the Government of India the large projects for irrigating the dry tracts of the central Deccan were postponed' to save risking any further losses on unprofitable regions.[156] In the famine year, the Government of Bombay made a case for investing to prevent far greater future expenditure, insisting that extensive irrigation was an 'absolute necessity' in the Deccan, because 'The problem of protecting the country from similar droughts in the future has become pressing.'[157] Temple proposed a local famine tax to raise money for further non-remunerative projects, but this was blocked by John Strachey, Finance Member of the Viceroy's Council, who was vigorously opposed to the irrigation lobby.[158]

While Temple's case for protective irrigation was convincing, the idea of charging the local population for it was extremely unpopular with local officers. The Collector of Khandesh argued, 'If local responsibility be strictly carried out, we should have to tell the famine-stricken people to pay for their own relief. This I think would fall little short of mockery.'[159] Moreover, the principle of district responsibility reversed an 1875 resolution that Local Funds were incapable of financing worthwhile projects, and that their resources should be aggregated for the Public Works Department to construct fewer but better works. The Ahmednagar Local Funds had indeed proved of little use, though partly because of interference from above. In 1874, two well projects were cancelled because the Bombay Government requisitioned the money set aside for them, including 1000 rupees for one in Kopargaon, and another was stopped half built when it seemed likely to exceed its meagre budget.[160] In 1876, the annual Local Fund budget for Shrigonda of 8,560 rupees set aside just 79 rupees for a water supply work, which was then decided to be 'advisable not to execute', because 'several wells built by Local Funds in previous years have tumbled in'.[161] During the famine, when resources were allocated for small irrigation works, they were spent entirely on cleaning, deepening and even 'excavating' existing wells; so many had fallen into disrepair.[162]

The notion of charging people extra for irrigation works was all the more reprehensible when it was well known that most *ryots* took little water only because they could not afford it.[163] In 1873, L. R. Ashburner, the Revenue Commissioner of the Northern Division, argued unsuccessfully that canal officers should not be responsible for overseeing the distribution of water, as they did not have cultivators' interests at heart.[164] When local famine funds failed to attract support, however, a similar system was devised to minimise losses from canal building. Under the 1878 Bombay Irrigation Bill, landholders close to government canals would be charged even if they did not take water, on the grounds that they benefited anyway from percolation through the soil. In fact, this had already been applied in Bhatodi village, where two *annas* per acre were added to the land revenue for the indirect advantages of the tank.[165] The Bill provoked extensive debate within the Governments of Bombay and India. The central Revenue, Agriculture and Commerce Department ambivalently opposed it, but were overruled. This prompted Hume, just six years before founding the Indian National Congress, into a private diatribe which reveals much about colonial decision-making, power politics and its sheer distance from the realities faced by peasants liable to pay what they could not afford for what they did not want:

> Now, the moment you get the compulsory rate, all check and control ceases. We are here in a close despotism. *Outside* the Government people may talk; but it is but as a murmur of the wind amongst the trees below. The Government sit aloft and care nothing for these things. *Inside* the Government, so long as the Financial Department and the Department of Public Works agree, no one else hears or knows or gets a chance of discussing the question. Once you have the compulsory rate, the sop has been thrown to the Triune Cerberus of the Financial Department. They are sure of their interest, and what more care they? Then hey for canals! Every engineer has a reputation to make; cost and returns signify now but little.[166]

After the Bill was passed, the Secretary of State sent a telegram objecting to the 'protection rate' clauses, on the grounds that there was no evidence or guarantee that those charged would gain anything.[167] The Government of Bombay did eventually lower all their water rates,[168] but were never able to reconcile the expense of establishing and managing irrigation works with the hidden nature of their value in arid zones,[169] and no more major projects were undertaken in Ahmednagar in the nineteenth century.

Conclusion

Conditions for cultivation were overwhelmingly difficult in Ahmednagar district from 1870 to 1884, and the area of cultivated land declined markedly during that time in extent and yields. The famine crisis of 1876–78 can at once be said to have been caused by such difficulties, to have created more, and also to have been part of them. It has been seen that the Government of Bombay – though aware of such hardships, and usually kept well informed by their local officers – interpreted evidence perversely, did little to alleviate problems and, at times, added to them. They did not care unduly what became of poor peasants, so they remained out of touch with their concerns, except during the declared famine, which they preferred to see as a unique period. Even then they did nothing to support cultivation by keeping peasants working on their land, as is seen in Chapter 5. When they did take an interest in agricultural production, it was with a blinkered insistence on profitable crops, which left the entire district marginalised and was often incompetently executed, or stymied by bureaucracy or financial constraints. Irrigation improvements were neglected and research into crop improvements singularly excluded foodgrains. The poor quality of the soil and the capriciousness of the rainfall were not, however, the only reasons why Ahmednagar peasants struggled to survive in this period. Price falls, exponentially rising debts and tax hikes formed a deadly combination from which they could not escape unaided. These are the subjects of the following chapters.

2
Market Opportunities, Risks and Failures

Introduction

This chapter examines the way in which agricultural markets affected peasant producers in Ahmednagar district, from the period of the cotton boom engendered by the American civil war, through the global recession of the 1870s to the period of famine crisis and its aftermath. An attempt is made to contrast the necessary conditions for market-led growth with those prevailing in the district at the time. The aim is to ask whether, in a poor dryland area, market forces helped or hindered smallholders. One of the effects of the famine crisis from 1876 to 1878 was an exponential rise in district foodgrain prices, from an average *jowar* price of one rupee three *annas* (1/16th of a rupee) per *maund* (80 *seers*) in 1873–74 to four rupees, two *annas* and six *pice* (1/12th of an *anna*) in 1877–78.[1] This common famine phenomenon was at once the cause of widespread entitlements failure and, according to contemporary economic logic, an incentive for private trade to bring in sufficient food to prevent starvation. Yet, despite the opening of relief works, 66 per cent excess mortality was recorded in Ahmednagar during the famine as well as extensive emigration.[2] Market forces were not entirely successful, therefore, in alleviating famine. But did they make it less – or more – likely to occur? Can this famine be linked, directly or indirectly, to the opening up of rural markets, or did it occur in spite of improved profits for some individuals?

This question arises because the 1876–78 famine took place in the context of a general attempt by the colonial state to improve the condition of western Indian agriculture by opening up wider markets, including the possibility of exports to Europe, thus bringing the profit motive into play. It was, therefore, a *laissez-faire* market approach to agricultural development. The benefits of such opportunities were apparent for districts able to produce cotton, whose markets were already integrated with the rest of Bombay Presidency. Ahmednagar, though not geographically remote from Bombay city, was on the periphery of the province in market terms. Transport and communications to the metropolis were not easy and it was of no particular economic importance. The Government of Bombay's agrarian policy did not, however, distinguish between districts. Ahmednagar's backwardness was recognised, but it was still anticipated that free market competition would do more to improve local cultivation than state investment or protection. If anything, this non-interventionist policy applied more to areas where the population and land were poor, because external investment in them was least likely to be profitable.

This created a chronic conflict of interest between the government and the primarily cautious, risk-minimising, subsistence-sustaining objectives of the mass of poor cultivators. Colonial conceptions of the peasant economy were essentially hostile to what they saw as the economic irrationality of farming choices that were not geared to increase profits, and therefore state revenue. Market exchange is, however, a double-edged sword, with inherent dangers as well as advantages. The power of the state allowed it to blur the distinction between a liberalising project designed to increase rural wealth and a punitive programme that forced many uncompetitive small farmers to sink or swim in the global market. The state's agricultural policies were designed to encourage the emergence of larger farmers, with a degree of capital, at the expense of small and indebted ones. This agenda was in itself optimistic in Ahmednagar and served to limit the degree of sympathy and support the British were willing to give to local peasants.

In order to investigate how this strategy affected Ahmednagar's poor peasants, it is necessary to consider their capacity to make profitable use of market opportunities. What are the material and organisational conditions for successful participation in competitive markets? In the following list, some suggestions are of absolute prerequisites, while others are ideal circumstances for export-led growth. All are relative. But if none were found to be present in an area, one might expect a strategy of local self-sufficiency to be more appropriate until they could be developed. First, it must be possible to produce and market goods for which there is a reliable demand, preferably an international one. To be competitive, producers will need quick and trustworthy information on the state of markets, and easy and inexpensive access to them, relative to their competitors. Therefore transport and communication infrastructure must be well established and within reach of all producers. Storage facilities would make it possible to benefit from market fluctuations. Production should be technically and allocatively efficient and preferably dynamic, generating the capital required for re-investment, expansion or methodological innovation. Producers should also be in ultimate control of all their economic choices, with mobile factors of production, notably capital and labour, enabling them to implement decisions quickly. In combination, capital and autonomy should be sufficient to permit the taking of risks – upon which all capitalist accumulation must be based – and to sustain losses, should some of those risks fail.

Ahmednagar's farmers were obviously not in an ideal situation to maximise global market opportunities. Peasant production of dry-crop foodgrains was far from dynamic. It could still be, however, that market solutions offered their best chance of improvement and this was the state's intent in demanding land revenue payments in cash rather than kind: to force peasants to trade more widely and to save profits both as an insurance against future bad seasons and as the basis for accumulation. This aim was undermined by the diversion of agricultural profits in Ahmednagar to trader-moneylenders, who were usually small. Though they were able to accumulate capital in the long term, *sowcars* initially borrowed from their urban kinsmen. Complex credit networks depended upon the circulation of money and therefore on the success of the entire presidency in global terms. Cheap foodgrains, on the other hand, were not traded widely, as they were grown primarily for consumption. Though some Ahmednagar grain was sold to nearby cotton districts, this was unusual, reflecting the activities of trader-moneylenders and the absence of other export commodities. Most grain remained within the district. Ahmednagar was not much integrated into wider markets during this period.

This chapter starts by examining local and regional cotton production and market opportunities, including the indirect benefits of selling foodgrains to cotton-growing areas. This moves on to a broader discussion of the extent to which Ahmednagar's markets were integrated with the rest of the presidency and how this affected local prices at a time of considerable local, regional and global market fluctuation. The paucity of transport facilities is examined before the state's optimistic market philosophies are contrasted with peasants' caution in the context of unreliable harvests, land revenue rises and moneylenders' control of the marketed surplus. Finally, the insistence from above that the state should not participate or intervene in the market during the famine crisis is shown to have been questioned by local officers in Ahmednagar.

Cotton production for export

Global demand for western Indian agricultural produce was limited to cotton, but strong enough to explain the Bombay Government's enthusiasm for export markets. In the 1830s, the Government of Bombay offered large advances to grow and supply cotton, and even exempted cotton fields from land revenue, though in Ahmednagar the unsuitability of the soil resulted in large losses to farmers.[3] In the 1860s, the outbreak of the American civil war and the embargo on southern cotton exports put a huge strain on the Lancashire mills, which translated into a major economic boom for Bombay growers, who diverted increasing amounts of land and resources into cotton. There was, therefore, export-led growth at that time, based on cash crop production, taking advantage of a historical opportunity in the hope of establishing a permanent share in the global raw cotton market. Cultivation figures suggest that the amount of land under cotton, even in the poorer Deccan, increased by 50 per cent in the 1860s, although in these early days of colonial statistic-taking, there was a tendency to exaggerate the impression of such increases.[4] However, the boom in cotton demand was not sustained after the end of the American civil war. Karl Marx argued that India's expansion and contraction of cotton cultivation in response to American supply amounted to exploitation of both producers and purchasers, rather than proving that the global market was self-regulating, as capitalists argued. Without a sustained European demand for Bombay cotton, the investment required for production at similar levels of quality to that from America never materialised.[5] In Ahmednagar, meanwhile, cotton cultivation remained insignificant, as can be seen in Table 2.1.

Ahmednagar was therefore well below the regional average in planting cotton and still further so in harvests per acre. Moreover, this was

Table 2.1 Cotton production, Southern Division, Bombay Presidency, 1869–70

District	Area of cultivation (acres)	Total yield (candies)	Yield per 100 acres (candies)
Ahmednagar	26,239	890	3.4
Poona	41,738	2,717	6.5
Satara	79,735	3,139	3.9
Sholapur	117,689	4,299	3.7
Dharwar	647,215	10,396	1.6
Belgaum	213,260	3,419	1.6
Kaladgi	597,822	6,008	1.0
Total	1,723,698	30,869	1.8
Entire Presidency	4,525,328	264,593	5.8

Source: General Report on the Administration of the Bombay Presidency (GRABP) (1870–71), pp. cxl–cxliii.

a relatively good year. Three years later, during a local drought, just 411 acres of cotton were cultivated in Ahmednagar.[6] While the districts broken down here are only those in the southern division of the presidency, the statistics are more impressive still for Khandesh, in the northern division. As a result of a government experiment, Khandesh cultivators planted foreign instead of indigenous strains of cotton, producing 56,564 *candies* (one *candy* equals 78 bales) of cotton – almost twice the total output from the entire southern division – from 540,808 acres.[7]

Knock-on benefits of the cotton boom

The failure to participate directly in the cotton boom need not have meant that Ahmednagar failed to benefit from the commercialisation of local agriculture. By switching to cotton, districts like Khandesh and Dharwar, with limited cultivable land, generated an increased demand for foodgrains from other districts. As foodgrains and cotton were spatial substitutes in fertile regions, one would expect them to be economic complements with high cross price elasticity. When Sir George Wingate introduced his revenue survey and settlement in the 1840s and 1850s, which played an important role in the commercialisation of Bombay agriculture, his intention was specifically that northern districts like Khandesh should maximise their comparative advantage in world markets, while the Deccan concentrated on foodgrains, including for export to cash crop districts.[8] Ahmednagar was said to be exporting up to 3.5 million tons of various agricultural produce *per annum* by the 1870s.[9] The repeal of wheat duties in January 1873 assisted this. It was the third most important crop in the district, well behind *jowar* and *bajri*, but much easier to sell. A small amount of valuable oil-seeds was also traded, and the area of irrigated land increased from its low base by an average of 5 per cent *per annum* over 30 years, helping to meet demand for better quality garden produce, sugar cane and grapes.[10]

As Banaji has pointed out, British satisfaction at the success of their commercial revolution contained a great deal of wishful thinking, and not a lot of reliable quantitative data. No disaggregated trade statistics were kept until the 1890s. Imports could be measured from municipality accounts of levies, but as no duties were charged on exports, records were not held.[11] Thus it was impossible to tell what proportion of imports was consumed and what was then re-exported, reducing the amount that Ahmednagar producers themselves had sold outside the district. Collector Boswell complained in 1876 that statistics were being demanded increasingly voraciously without sufficient time to collate them, adding 'it is to be feared that when prepared hurriedly they are very unreliable and the possession of a mass of inaccurate statistics is of questionable benefit, as persons are wont to accept as true whatever they find in print or in Government Returns'.[12] The need for improved accuracy in agricultural data was recognised by the Bombay Government,[13] but this did not prevent them from interpreting what they had as they chose. Trade figures in particular were not routinely kept, leaving Boswell confident in his view that Ahmednagar trade was very small compared to other districts, but unable to prove it. When sufficiently disaggregated trade statistics became available, Banaji gives the example of the 1891–92 Bombay Administration Report recording that 'jowar was largely exported abroad', when the figure was calculable at only 3.8 per cent of total yield.[14] Given that exports were generally higher by then, there is no reason to suppose that similar assertions in the 1860s or 1870s were any more accurate.

Cheaper grains did increase in value, temporarily, in response to cotton profits at the height of the American civil war. This inflation was caused, however, by increased local money supply, rather than demand for a greater volume of foodgrains in the presidency as a whole.[15] They could be exported to other districts, but such trade was unreliable, reflecting the global markets which cotton producers served. This was acknowledged by the Survey Commissioner, Colonel Anderson, when assessing part of Rahuri *taluka*,

> The almost exclusive cultivation in these villages of the common food grains, which are subject to so great fluctuations in price, is a disadvantage which...is entailed to a great extent by the quality of the greater proportion of the soil, which is not suitable for the growth of the higher class of products in constant demand at steadier prices for exportation.[16]

Prices could potentially fall either through poor cotton harvests, reducing wealth and therefore demand or, conversely, through the acquisition of higher tastes in successful areas. Although the period of the American civil war did see some benefit for Ahmednagar in the form of foodgrain prices and greater availability of money, the volatility of that market, the low quality of the commodity, and the determinedly local nature of demand, all meant that the boom period was much less remunerative and sustainable there than elsewhere. In the critical 1870s and early 1880s, the import of foodgrains for local consumption was consistently noted, rather than their export.[17]

There was another way, however, in which Ahmednagar might have gained by the profitable cultivation of cotton for world markets. Considerable quantities passed through the district, heading for the Bombay ports from the north and east – Central Provinces, Berar and the *Nizam*'s Dominions. This generated trading income and hire charges during the summer for those peasants who owned carts. It also provided state revenue, in the form of import duties, as well as road tolls and ferry fees. These receipts went into Local Funds to be re-invested in further road construction or repair, as well as other small public works like tanks or *dharmsalas* (rest houses). If through-trade encouraged and financed the development of internal communication networks within Ahmednagar, it could only help local producers. However, the extent of such improvements was limited, as is seen later.

Nonetheless, an indirect share in the 1860s boom could be had from the passage of cotton through the district, and this might have helped to mitigate poorer trade years in the 1870s, when cotton and food prices both dropped. When cotton traffic slowed, however, Ahmednagar was left at a disadvantage by its lack of a railway – until the construction of the line from Dhond, in Poona district, to Manmad, in Nasik, by workers on famine relief in 1877–78. As this ran from north to south, it was no help in exporting produce to Bombay city to the west, but it nonetheless aided the diffusion of agricultural and other goods from north and south India, and re-established Ahmednagar as a through-trading centre. The British, who saw the district's trading location as its chief comparative advantage, encouraged this role. In 1879, Collector Stewart initiated the construction of a cotton green for trade and storage, near the newly built station.[18] These two developments also inspired a modicum of local industry in the form of three cotton presses, although one had closed down by 1881.[19]

Whatever trading profits it may have conferred, the green did not affect local cotton production. In 1878–79, cotton was sown in 26,881 acres – almost exactly the same as 1869–70, as Table 2.1 shows.[20] This was a long time, moreover, after the inflated world demand during the American civil war, which had ended in 1865. As the intervening period contained a long economic slump and a famine, the failure to bridge the gap sooner needs to be considered. This state investment in Ahmednagar as a transit market for external produce also confirms that the commercialisation of

Bombay's agricultural markets was focused upon valuable export goods rather than foodgrains, for which little was done to increase local trade.

The degree of integration of Ahmednagar with external markets

The issue of how local production was affected by wider markets is a paradoxical one. On the one hand, it is possible to argue that Ahmednagar farmers did not benefit from the globalisation of agricultural markets because it did not really extend to the cheap foodgrains they produced. On the other hand, it could be held that *jowar* and *bajri* markets did develop on the back of the commercialisation of exportable products – but that exposure to the vagaries of wider market prices proved harmful by reducing security. Examination of price fluctuations within the district can reveal whether either, or both, of these possibilities can be substantiated. Table 2.2 shows the annual average prices for *jowar* in Ahmednagar city from 1867 to 1883. The figures given are for pounds of grain per rupee – that is to say, the higher the figure, the lower the prevailing price – with the final column calculating the cost of a hundredweight of *jowar* in rupees, *annas* and *pice*.

Table 2.2 Annual average *jowar* prices, Ahmednagar city

Year	Price (lbs per rupee)	Rupees per 100 lbs
1867–68	45	2-3-7
1868–69	25	4-0-0
1869–70	30	3-5-4
1870–71	45	2-3-7
1871–72	32	3-2-0
1872–73	41	2-7-0
1873–74	67	1-7-11
1874–75	81	1-3-9
1875–76	66	1-8-3
1876–77	34	2-15-1
1877–78	19	5-4-3
1878–79	21	4-12-2
1879–80	23	4-5-7
1880–81	41	2-7-0
1881–82	72	1-6-3
1882–83	56	1-12-7

Source: *Ahmadnagar Gazetteer*, p. 555.

As these are annual averages, they give no indication of intra-year price fluctuations, which could be very erratic, especially in poorer seasons. However, they confirm the volatility of district foodgrain markets, suggesting a lack of integration with wider markets and low volumes of external trade in either direction. By implication, local traders lacked the capital, information and infrastructure to respond to price variations, making the Ahmednagar grain market uncompetitive and liable to acute failure. Only local demand could be counted on, making it inelastic in spite of the elasticity of the rain-dependent supply. The years in which prices rose considerably – 1868–69, 1871–72 and most of all 1876–78 – correlated directly to seasons of local scarcity, implying that they related primarily to local demand. The famine affected most of South India, but the 1871–72 shortage, for example, was restricted to Ahmednagar and Nasik to the north. By contrast, prices fell in 1873–74, when there was a good local crop, despite famine in Bihar, leading Boswell to remark:

> It is strange that in a season when there has been such an extraordinary demand for grain in some parts of India it should be so exceptionally cheap through the whole season in this district and that too when it is not so in the neighbouring districts of Poona and Nassick. This fact shows strongly the isolation of the district which... points to the necessity of Government doing more to improve the outlets for the produce of this district if it is to maintain even its present degree of prosperity.[21]

Wholesale produce prices were almost always lower in Ahmednagar than Poona – and in the Deccan than Bombay city – and the difference was particularly marked at a time when the district had surpluses for potential export. Table 2.3 shows the prices in 1873–74 and at the height of the famine, per standard *maund* of 80 *seers* (just over two pounds).

Table 2.3 Annual average prices for *bajri* and *jowar* in Ahmednagar, Poona and Bombay markets, 1873–74 and 1877–78 (rupees, *annas* and *pice* per *maund*)

Market	1873–74		1877–78	
	Bajri	Jowar	Bajri	Jowar
Ahmednagar	1-3-11	1-3-0	4-2-5	4-2-6
Poona	2-1-0	1-8-5	4-11-5	4-6-8
Bombay city	2-8-1	2-1-1	5-3-0	4-7-10

Source: GRABP (1873–74), pp. cviii–cix and GRABP (1877–78), pp. cxvi–cxvii.

If prices therefore reflected local supply, because of low levels of external trade, the aggregate result for producers was income equilibrium. When the harvest was good, they could not get good prices, and when prices were high, it was because production had been low. This did not mean that variable rainfall made no difference, however. Whereas income did not vary much according to the size of the foodgrain output per season, food security did. Many producers in net deficit had to turn to the market for their own consumption needs, making high prices a disadvantage which could exacerbate, or precipitate, indebtedness. Increased demand from farmers whose crops had failed then raised prices further, inducing effective demand failure. In extended scarcities, this could cause entitlements failures. As Sen has shown, purchasers unable to afford prevailing food prices become famine victims if prices remain high for sustained periods, regardless of the total volume of food in the market.[22] Moreover, rain failure was never even, and some producers could lose their entire produce, even in relatively average seasons, as seen in Chapter 1.

On the other hand, the land revenue demand, though calculated on the basis of an average of good and bad seasons, was more likely to be reduced, suspended or remitted in cases of hardship following poorer seasons. Under the *annewari* system, full payment would be required when grain prices were lowest due to local surpluses. As early as 1848, assessment reports had recorded *ryots* selling their bullocks to pay the revenue after large harvests.[23] Moreover, revenue pressure, by deliberately encouraging increased production – at diminishing rates of return, on worse and worse land – also exacerbated a general downward tendency in foodgrain prices before the famine. This raises a further anomalous attitude on the part of the British towards foodgrain producers. The effects of supply and demand within a relatively small climatic zone meant that the size of production did not greatly increase the overall income of producers. Yet better seasons were applauded (and often exaggerated) in annual reports.[24] This might suggest a greater concern for consumers than for poor grain producers, or just contentment at greater revenue collection. However, peasants' welfare was also enhanced by good rainfall, because it averted the risk of calamity for that year. Thus the celebration of good seasons rather reflected the extent to which both local state discourse and peasant aspirations revolved around the constant fear of the costs of major crop failure.

Yet the state failed to assist peasants' own attempts to minimise the effects of that eventuality. The combination of capricious rainfall and market forces meant that household resources were always limited, but the government demanded its share. Thus while prices fluctuated wildly according to local supply and demand, state revenue policy obliged peasant producers to rely upon an unintegrated market as if it were competitive. Local production was reported to have largely determined price levels well into the 1870s. However, Table 2.2 also shows that grain prices fell rapidly between 1872 and 1875, which was not wholly explained by reasonable district harvests, particularly as the same period saw the nearly 50,000 acre drop in the cultivated area noted in Chapter 1. It would appear, then, that global influences did affect Ahmednagar after all. Following the end of the American civil war, Bombay Presidency suffered a marked slowing of export trade. The Bank of Bombay, which had emerged and grown rich during the frenzy of cotton speculation in 1863, and underwritten several large projects, including the reclamation of Back Bay in Bombay city, collapsed under the weight of bad debts as early as 1868.[25] By 1870, cotton cultivation had fallen back dramatically and declining cultivation in Ahmednagar reflected this slump. Nor was this just a regional problem. Having participated so successfully in global markets in the 1860s, Bombay cultivators and traders found themselves tied to the coat tails of the first major worldwide recession, lasting from 1873 to 1879. This naturally had important consequences in Europe, signalling the end of the so-called 'era of free trade', but western India was less well equipped to cope. To make matters worse, in 1876, the devaluation of silver against gold in world bullion markets, and consequently of the rupee against sterling, increased the costs of colonial administration in real terms, and imposed a new mood of stringency upon a provincial government which had hitherto been renowned for its profligacy, if not necessarily for its generosity towards its subjects.

Judging by the timing of the fall in *jowar* prices in Ahmednagar, grain producers were among the first to experience the effects of this global slump, despite having benefited last and least from the boom. This assumes that local price falls were indeed induced by global price trends, which was the common view at the time. The Government of Bombay admitted in 1874 that prices had fallen even in districts where there had not been a good season. 'A fall so general cannot, therefore, be due to any such temporary cause. There can be no doubt that it has resulted, not from an abundance of agricultural produce, but from the scarcity of money.'[26] Thus, though Ahmednagar grain markets were scarcely integrated with the international cotton market, the two were connected by regional money markets. There was no shift in terms of trade to the advantage of cheaper foodgrains. Rather, as the world economy sank into depression, those recently thrown into it at the bottom were crushed deeper into the mire by its weight. With both cotton and foodgrain producers facing entitlements failure, there was a pan-regional lowering of effective demand, creating the appearance of an integrated price slump across unintegrated markets. Moreover, if prices fell in years when yields were low, cultivators suffered twice over, dramatically reducing their income.

The government recognised that the demand for *jowar* and *bajri* had collapsed far more severely than that for cotton noting that

> whilst the people of the coast and the cotton districts, being able to continue growing produce capable of profitable sale even at low prices, have been able to meet the more pressing and dangerous of their liabilities on comparatively favourable terms, the inhabitants of the Deccan have been unable to dispose of grain, with which the markets are glutted.[27]

As a result, foodgrain cultivation on arid soil was admitted to be unviable: 'the cheap prices at which the commoner grains are now selling do not leave to the farmer of poor land a sufficient surplus, after the Government share and expenses of tillage have been paid, for the maintenance of himself, his family, and his cattle'.[28] Despite this, when the district's 30-year revenue settlement came up for review in the late 1870s, the Bombay Survey Department – ignoring calls for caution both from district officers and the Government of India – based its revisions more on the average increase in prices over the full 30 years, including the exogenous boom of the early 1860s, than on the sharp downward trend at the time of assessment. The push to profit in order to pay the revenue – and earn the continued right to cultivate government lands – was therefore magnified just when opportunities to profit, and also to borrow, were worse than they had ever been.

It would appear, then, that being linked to a wild trade cycle of growth and recession was even worse for small peasants than the trading disadvantages of market isolation. Ahmednagar appeared to suffer a worst-case combination of the two, with external demand for its produce insufficient to guarantee a steady profit, but links to outside markets strong enough to see the knock-on effects of wider deflationary pressures. Assistant Collector Loch suggested that exogenous price – and therefore profit – falls were the chief reason why cultivation drastically declined in the district in the early 1870s, and also why so few peasants were able to build wells.[29] It is possible to corroborate the impression that Ahmednagar markets were affected by those in Bombay Presidency generally, without being integrated into them, by examination of government stamp fee receipts, which reflected the level of commercial activity. Table 2.4 gives the figures from 1861–72 for both Ahmednagar and Bombay Presidency (including Sind, the Native States overseen by the British and also Aden).

Table 2.4 Net stamp receipts (rupees), 1861–72

Year	Ahmednagar	Fluctuation from base of 100%	Bombay Presidency	Fluctuation from base of 100%
1861–62	301,927	100	2,947,583	100
1862–63	219,380	73	3,175,324	108
1863–64	262,301	87	4,339,191	147
1864–65	316,517	105	5,540,468	188
1865–66	318,758	106	4,698,146	159
1866–67	300,008	99	3,557,402	121
1867–68	372,409	123	4,304,366	146
1868–69	385,078	128	4,486,167	152
1869–70	289,353	96	4,460,996	151
1870–71	260,726	86	5,321,092	181
1871–72	221,621	73	4,769,616	162

Source: GRABP (1871–72), p. 138.

Care needs to be taken with these figures, which included fees from civil courts. Up to 65 per cent of stamps derived from suits for indebtedness. However, credit tended to expand and contract according to the volume of trade. Overall, a clear pattern can be discerned. First, the standard deviation is noticeably less in Ahmednagar than in the presidency as a whole and, in particular, the high points of stamp revenue are less above average in the district. Otherwise, there is a fair degree of synchronicity between the two columns, with Ahmednagar trade picking up with the start of the cotton boom in 1863. Once the wider recession starts to loom, however, Ahmednagar's trade drops far more sharply and continually than the presidency's does. Throughout the period, and especially at its end, Bombay did better than it had before the American civil war, while Ahmednagar's position declined in the 1870s.

Following the stamp receipts in Ahmednagar through the next decade, the steady decline of trade and credit in the build-up to the famine crisis becomes obvious. At this point, figures differentiating between civil court and general stamp receipts (reflecting the number of business transactions) became available, and these are shown in Table 2.5.[30]

Table 2.5 Ahmednagar stamp receipts (rupees), 1871–81

Year	Total stamps	100% as base	Civil court stamps	100% as base	General stamps	100% as base
1871–72	236,034	100	138,805	100	97,229	100
1872–73	256,803	109	166,144	120	90,659	93
1873–74	223,235	95	147,046	106	76,189	78
1874–75	204,331	87	138,288	100	66,043	68
1875–76	195,626	83	135,693	98	59,933	62
1876–77	175,491	74	115,453	83	60,038	62
1877–78	152,877	65	89,198	64	63,679	65
1878–79	174,199	74	103,563	75	70,636	73
1879–80	143,260	61	79,338	57	63,922	66
1880–81	99,690	42	55,360	40	44,330	46

Sources: GRABP (1871–72), p. 136; GRABP (1872–73), p. 487; Boswell to Havelock, No. 1645 of 1874, 20 July 1874, p. 131; Boswell to Oliphant, No. 2132, 20 July 1875, p. 480; Boswell to Havelock, No. 1952, 20 July 1876, p. 243; Jacomb to Robertson, No. A/4960, 19 July 1877, p. 37; Stewart to Robertson, No. 3195, 22–4 July 1878, p. 315; King to Robertson, No. 3140, 19–23 July 1879, p. 42; King to Robertson, No. 4161, 20 July 1880, pp. 54–5; King to Robertson, No. 4584, 22–25 July 1881, p. 115.

While civil court cases increased in the aftermath of the 1871–72 scarcity, and both columns picked up slightly directly after the famine, the general trend is strongly downwards. Court and trade stamp receipts broadly echo each other, though it is striking that trade receipts fell consistently from 1868–69, apart from the famine period itself, when they were held up by imports, whereas credit dealings contracted only when the famine crisis emerged. The Deccan Riots of 1875 affected both and the effects of the 1879 Deccan Agriculturists' Relief Act were almost as marked that year on general stamp income as on the numbers of petitions by moneylenders to the civil courts.[31] Even given this, the continued trade decline after a very brief post-famine recovery contradicts local officers' optimistic assessments of peasants' welfare at that time. In 1880, Collector King claimed there was 'a very fair share of material comfort still left to the ryot who is willing to live the simple frugal life which sufficed his predecessors'.[32] As well as being contradictory and implausible, this observation effectively endorsed long-standing peasant conservatism.

Transport and communication infrastructure

The price statistics published by the British (and used when calculating a district's wealth, for example for revenue purposes) were, as in

Tables 2.2 and 2.3, based on selling prices in local markets. The value of produce in Ahmednagar town was naturally higher, though, than in the *mofussil* and reflected the price obtained by local traders rather than peasants themselves, who commonly dealt with village *sowcars* in kind. Famine Commissioner Mahadev Barve argued that they engaged in no trade of their own and had no access to price information, and as a result paid inflated rates for oil, spices and tobacco because they did not realise the relative value of their grain.[33] This raises another criterion for successful participation in global markets mentioned at the start – that of transport and communication infrastructure. In proclaiming the benefits of wider markets to the agricultural sector, the colonial state made much of the ease with which private traders could import or export produce in order to take advantage of better prices, either within a district or abroad. While grain prices were published, it is questionable how widely they were disseminated, except during famines. Even peasants who controlled their own sales were unlikely to have been aware of higher prices outside the district. Moreover, if prices were only published for one place within a district, more local gluts or scarcities – such as that in Kopargaon *taluka* in 1881 – could not expect much response from the market or the state itself.

Information was in any case of little use either to profit or to benefit from imports, unless remote areas – and whole districts – were adequately served by road or rail. Successful market policies usually require prior state investment in the construction of transport facilities to ease the passage of free trade. Only in Britain was a complete railway network built on unassisted private capital. In western India, major lines such as the Great Indian Peninsula and Bombay, Baroda and Central India Railways were built by private companies, but remained under the administration of the Government of Bombay. It was no surprise that venture capitalists were not interested in providing transport to a backward district like Ahmednagar, and no line was ever built from there to Bombay. Indeed Boswell warned in 1874 that the price discrepancies between Ahmednagar and neighbouring districts was being accentuated by the opening of the Great Indian Peninsula line nearby, leaving the district less able to participate in regional markets than ever.[34] Its own first line, from Dhond to Manmad, was built in 1877–78 by the government itself and, like most in the Deccan, never made more than the slightest profit.[35] The aim of the Dhond–Manmad line was not to give Ahmednagar's grain producers access to more profitable markets, however, but to assist national through-traffic. Until it was built, all trade from north to south India had to go via Bombay, involving an arduous and time-consuming trek up and down the Western *Ghats* (hills).[36] The fact that it went through Ahmednagar at all was coincidence, as the Bombay Government believed that the Deccan had a sufficient rail network. There was also no line in the famine hit Southern Maratha Country, but the Dhond–Manmad line took precedence, as a commercial route.[37] Priorities were blurred further by Superintending Engineer Major-General St Clair Wilkins, who argued for further lines from 'a political and military point of view' first, and 'in a commercial and famine aspect' only second.[38]

As seen in Chapter 1, the Governor of Bombay, Richard Temple, did argue, without support, for state investment to make *laissez-faire* work, especially as a famine policy, arguing for extra taxation if necessary.[39] That he should do so after the famine undermined his claim that grain imports had met the demand using existing facilities.[40] The question of trade during the famine is considered later in this chapter. Assistance to potential exporters was even more easily overlooked. King revealed in 1881 just how much the railway had been planned to help local trade:

> I hope I may be pardoned if I say that the policy which determined the line along which the railway has been traced, though no doubt justified by important consideration, has not succeeded in making the railway so tangible and accessible a benefit to the public generally as it would have done, if the existence of large towns and villages had attracted more notice and had more directly influenced the selection of a route.[41]

In other words, in its planners' eagerness to find the cheapest way possible through the district, local markets had been ignored. In addition, many cultivators had been forced off their land to make way for the railway. Although Assistant Collector Candy usually negotiated privately, rather than using the terms of the Land Acquisition Act, which limited sums payable, his priority was ordered by the Government of India to be 'the greatest economy in transaction'.[42]

In the initial enthusiasm, in 1878, to maximise the advantages of the new railway to aid recovery from the famine, Stewart declared that it was, 'likely to give a great stimulus to the markets and trade of this district'. Yet, he admitted, 'a great deal remains to be done in bridging rivers and nullas and rendering the line workable during the rainy season'.[43] Not until 1880 was the line fit for traffic throughout the year.[44] Moreover, Stewart, as has been seen, was chiefly preoccupied with the benefits of the through-trade in cotton. Feeder roads, to link local markets to the railhead, were immediately proposed, but depended on Local Fund resources which had been bled dry by famine relief projects. An appeal by Stewart for provincial funds for these works received no response.[45] During the famine itself, they had not been constructed because 'roads are stated not to be needed for relief'.[46] Thus, given financial constraints, local market interests were subsumed both to immediate relief and to wider trading concerns. All sides agreed their necessity, but not their priority. The first reports of construction work on feeder roads were not despatched until 1882. This was ironically undertaken once more by workers on relief in Kopargaon – which had been predicted to benefit particularly from the railway.[47]

Improved transport can have a dynamic impact on society, particularly by creating access to markets with the potential to generate either profit or food security. It can also cause difficulties by integrating local markets with some wider ones in which they are not competitive, including that in transport itself, but not with others of greater potential benefit. So it proved with the arrival of the railway through Ahmednagar, which not only failed to improve local access to grain markets, but also had some negative impacts upon the district. Those *ryots* who had previously been able to supplement their income by hiring out their carts suffered particularly. In 1876–77, they could command an average of two rupees *per diem* in Ahmednagar town.[48] The following year, with famine relief still operating and thus extra grain supplies to carry, they had fallen to one rupee, eight *annas* and five *pice*, which Stewart explained by the opening of the railway.[49] From 1878–80, well before grain prices themselves had returned to pre-famine levels, a cart could be hired for an average of just one rupee per day.[50] This price was similar to that before the famine, but represented a lost chance of compensation for low yields, and a long term threat to hauliers' livelihoods. King suggested that continuing cart traffic, direct to Bombay, would last only until the benefits of the railway were 'fully recognised', but later acknowledged the harm to cart owners.[51] Again, then, it would appear that the advantages of the railway conflicted with the interests of the local population, but that the British believed such problems to be of lesser significance. That some long-distance cart traffic continued suggests either mistrust of the railway, its failure to carry enough grain or willingness by small cart owners to continue working for less than a subsistence wage – all of which were disregarded by officers wishing to see the Dhond–Manmad line as a panacea.

Table 2.6 shows cart numbers in the district, according to village returns. It should be noted that these returns, which only started to be

Table 2.6 Cart totals in Ahmednagar district, 1876–80

Year	Carts
1876	25,053
1877	21,447
1878	22,264
1879	21,360
1880	21,802

Source: King to Robertson, No. 4584, 22–5 July 1881, p. 49.

kept in the 1870s, were the same ones that recorded livestock numbers which were, as seen in Chapter 1, noted for their unreliability. Assistant Collector Woodburn reported in 1876 that Shrigonda *taluka* figures were exaggerated in an attempt to fit reality to a prescribed form with columns for two and four bullock ploughs. 'I find that often when a Kunbi has one plough "with eight bullocks" the Koolkarni registers him as having *two* ploughs "with four bullocks".'[52] A similar miscalculation for carts was likely. What is more, an extra appendix in 1877 revealed that, of the high 1876 total (village returns always appeared a year late in annual administration reports), 9,270 were 'riding carts', with only 15,333 able to carry loads.[53]

It might be argued that the large drop in cart numbers in 1877 reflected less accurate counting during the chaos of famine, but Stewart and Revenue Commissioner Robertson believed emigration to be the chief explanation.[54] Famine migration from Ahmednagar was considerable, as will be seen in Chapter 5. In general, inaccuracies tended to exaggerate rather than underestimate totals, yet numbers remained low after the famine. The further large drop in 1878–79, despite the logical anticipation of recovery, suggests a negative impact of the railway upon cart trade. Whereas this effect on long-distance carriers was predictable, the overall decline in carts is perhaps surprising, given the lack of proximity of the railway to local markets. One might expect business between the two to be very lucrative, giving cultivators who also owned carts particular market advantages, as in Neil Charlesworth's theory of the emergence of rich peasants.[55] That this does not appear to have happened reinforces the impression that the Dhond–Manmad railway did not carry a significant amount of Ahmednagar grain produce.

Another group of people adversely affected by the railway was the *Mhar*, *Ramosi* and *Bhil* men who had previously acted as watchmen and guides for travelling traders on local roads. King's proposal that they might replace imported labourers from the Upper Provinces in performing menial tasks on the railway itself was rejected, no doubt not helped by his warning that, without their old jobs, they were liable to resort to crime.[56] Robertson had little sympathy for their cause, declaring, 'unless the laws of demand and supply are acting perversely in the case of these people, it is no great matter of difficulty for them, if so inclined, to accommodate themselves to the times and to find employment in the larger towns and villages'.[57] But labour demand was indeed poor and resettling in unfamiliar circumstances was expensive and fraught with social difficulties.

Cultivators generally lost by the arrival of the railway, then, especially in poor seasons when prices were also low, because it created more opportunities to import grain than to export it. The notion that opening the market could reduce local food insecurity was less important than the supposed opportunity for external profit, until the famine itself, when free trade was credited to an unrealistic degree with meeting all of the demand for food. Afterwards, when Ahmednagar's foodgrain markets had become somewhat more integrated with those around them, grain did arrive in respectable quantities to meet the demand when it was large enough, as in the local scarcity of 1881–82. Ironically, even then, Collector Elphinston recorded that grain poured in, 'from the Nizam's territory by the two Provincial roads to Toka & Pythan and by the Kharda Kashti road (which tho' a mere track carries very heavy traffic)' over and above the volume which arrived from the north by train.[58] This did not necessarily constitute an improvement anyway, because the arrival of the railway had also discouraged local traders from storing grain within the district.[59] Moreover, 1881–82 was not a total crop failure. Imports primarily reflected surplus production in other regions, rather than a response to prices in Ahmednagar. By flooding the market, outside traders benefited consumers in the district but exacerbated the hardship of district farmers who did have some grain to sell. Elphinston compared the situation to the decline of English smallholders due to cheap grain imports, which was about as close as a British Officer could get to criticising free trade in 1882.[60]

Though traders, like peasants, tended to act conservatively, they did not face ruination in the event of famine, and were somewhat less risk-averse as a result. The decline in local storage after the arrival of the Dhond–Manmad railway was a good example. Storage was always difficult, and few district *banias*, let alone peasants, could achieve economies of scale sufficient to justify the expense of maintaining grain pits in order to benefit from off-season prices. Improved transport made it less important – and relatively less profitable – to keep stocks locally for sale. This threatened Ahmednagar's food security, just as it had in better off districts during the famine itself.[61] In their official dissent in the Famine Commission Report, commissioners H. E. Sullivan and agricultural economist Sir James Caird suggested that the government should store grain itself, arguing that private trade could not logically be expected to supply all sudden food needs.[62] This notion was dismissed as an unwanted intervention in the grain market, even after private traders had closed their pits.

If the railway proved a mixed blessing when it was finally constructed, what of existing transport facilities within the district, or leading out of it, which could have helped increase trade in local produce before the famine? Road building could be undertaken, as was the case with the railway feeder proposals, from provincial, imperial or local funds. In Ahmednagar, very few road projects were executed under the former heads throughout the decades before and after the famine. They were costly in terms of European manpower in the PWD, and notoriously slow. For example, there were no properly surfaced roads leading into the remote and hilly *taluka* of Akola before 1877, and a provincial project to open it to trade between Loni and Bari was far from completion. As a result, cultivators missed out on preferential rates below the *ghats*, and had to survive in the static local markets, which did not earn them enough to stay out of debt until the next harvest. When Assistant Collector Hamilton requested 10,000 rupees to finish the road as a relief work, Collector Jacomb responded that they ought to be grateful for the money already spent on the road, thus completely evading the question of its failure to serve its intended purpose.[63] Only when Hamilton protested that he had instead been given money to build feeder tracks to a route which would thus be left unpassable did the Bombay Government relent.[64] Similarly, Kharda, the grain market in Jamkhed *taluka* in the southeast remained cut off from its key trade to the *Nizam's* Dominions, because a hill road towards the railway would have been too costly, obliging produce to be carried overland on pack bullocks. Woodburn's call for a new track in 1877 was not heeded, leaving the region isolated even after the famine.[65] Even the main road from Dhond to Manmad, already a major thoroughfare before the railway, was reported in 1875 to be 'considerably cut up by the large cotton traffic which passes over it in the hot weather' and thus unpassable after the monsoon had broken, without repair.[66] The following year it was pointed out that the lack of bridges over *nullas* (streams) and riverbeds added hugely to the length and difficulty of journeys on the same road.[67]

Local Fund works, conducted from the district's own resources and overseen by revenue officers, were often pitiful in quality. As Local Funds derived partly from through-trade, in the form of *octroi* (freight tax) and tolls, the decline already in place in the 1870s made matters worse. In an attempt to overcome this, the Bombay Government altered the system in 1875, transferring Local Funds to the professional Public Works Department for execution. This inevitably meant that still fewer improvements were made, with Boswell complaining that the effect was 'to raise the cost of construction of many works 25 to 50 per cent'.[68] The consequent loss of a district engineer in the interests of economy also, he suggested, nullified the advantages of the Local Fund system, which was left no better able to provide viable facilities for local trade than the tightly watched provincial purse.[69] However, he also complained that Local Funds were spent inefficiently,[70] thus undermining his case for more roads by allowing his superiors to revert to the imperial priority of economy.

Roads built as a result of famine relief works also did little to improve the district. Attempts to keep relief spending down, which are discussed in Chapter 5, meant they were of even lower quality than usual. Among many examples, after inspecting work on an important route from Rahuri to Belapur, Robertson reported that 'I do not anticipate that much of the road will be found after one monsoon.' In an echo of the earlier criticism of Boswell for calling for more cheap local projects, he added, 'it is far better to spend twice or three times the sum on one good road under the PWD than waste the whole money in dribbles on several roads under Mamlatdars'.[71] Jacomb did not disagree, reporting that busy Local Fund Committees only found time to inspect roads after completion, when it was too late to eradicate common flaws in construction.[72] However, when projects were proposed, any with large estimated costs like the Loni–Bari road in Akola were automatically rejected, and only cheap and less important works sanctioned. The Dhond–Manmad railway took most of the PWD famine workforce anyway, with road works kept to a minimum.[73] The Government went as far as to warn local officers that 'care must be taken not to over-rate the value of lines of communication'.[74] Yet Robertson warned that smaller famine works had 'borne so heavily on Local Funds that it will be long before they can recover and do much good in the country'.[75] In addition, Assistant Collectors' discretionary allowances for road works were blocked to save more money.[76] Temple's concern to improve communication after the famine was not going to affect roads in Ahmednagar for some time.

After the famine, Colonel Jenkin Jones of the PWD noted that, 'Considering the size and importance of the district, it must be said that road communication is very defective.'[77] At this point, collectors were instructed to record total communication facilities, suggesting that the link between market access and famine was understood. In 1879, there were 122.25 miles of railway in Ahmednagar, just 31 miles of wholly bridged roads, 270.5 miles partially bridged, 59 miles with no bridges but surfaced and 22 miles suitable only for fair weather.[78] Thus at the height of the crusade to open agricultural markets, Ahmednagar producers often had no reliable outlets for their grain at all – and even the construction of a major railway through the district did relatively little to improve the situation.

Peasant poverty and rational caution

With low productivity, efficiency and demand for the poor products they harvested, and few transport facilities to market them, Ahmednagar *ryots* were thus in little position to benefit from global trade. Moreover, the cultivating population of the district was chronically under-capitalised. Not only were they poor, many peasants had little access to cash. When they did have any wealth saved, for example from their wives' dowries, they tended to keep it in the form of gold or jewellery, which could be sold in times of crisis, and therefore acted as insurance. The poverty of the majority of the population made them risk-averse, in the sense that they saved against emergencies rather than re-investing in the land, though they were not necessarily able to do so sufficiently to survive the famine. Gluts in Bombay gold markets in the early 1870s suggested that large volumes of such ornaments had been passing steadily out of the peasants' hands for some years.[79] In the famine, therefore, assets such as agricultural tools, carts and livestock had to be sold as a coping strategy, despite their importance for production and the well-established preference of peasants for maintaining their livelihoods.[80] Moreover, while male peasants saw jewellery as the obvious saleable household asset, being both non-productive and prone to less severe downturns in terms of trade, Agarwal has pointed out that this could significantly increase women's vulnerability. Jewellery was often the only asset that women owned within households. If disposed of early in times of crisis, they were left with nothing to fall back on if later separated from their husbands on relief or through individual migration or death.[81]

In theory there remained the possibility of a peasant elite emerging, perhaps from existing socially pre-eminent groups, with control over its own – and others' – trade arrangements that would enable it to make money and become a capitalist class. Such peasants could then re-invest, not only in land improvements like wells, but also in expansion of their holdings at the expense of their neighbours, using credit as an instrument of acquisition. Charlesworth has argued that this was indeed what eventually happened in the Bombay Deccan, albeit slightly later, after the passage of the 1879 Deccan Agriculturists' Relief Act. This legislation, which is discussed in Chapter 3, penalised professional moneylenders and so gave the opportunity to a big peasant cum *bania* cum moneylender group with a greater tendency than *Marwaris* to covet other cultivators' land.[82] However, Charlesworth bases this conclusion primarily on evidence from Satara district, which, though it contained some areas of worse land than Ahmednagar, also included some more fertile pockets conducive to profitable cultivation. There is little evidence of the emergence of a big peasantry in Ahmednagar, during or after the period in question, at least until a renewed attempt was finally made to provide state irrigation in the 1910s. This gave rise to notionally co-operative sugar cultivation, which was indeed dominated by semi-capitalist elites, although even then not all derived from the local peasantry.

The final criteria for successful participation in global trade listed earlier were that producers should be in ultimate control of all their economic decisions and able to take risks and sustain losses. The social structure of trade in the Deccan prevented this, because of widespread agricultural indebtedness. In the absence of substantial farmers looking to borrow to expand into profitable commercial agriculture, rural trader-moneylenders in Ahmednagar sought to lend to struggling dryland cultivators instead. They not only extracted the surplus value of cultivation but also effectively made *ryots*' farming decisions for them, by giving advances in the form of seeds and implements. Lenders also controlled local grain markets. Interlocked credit contracts allowed them to take the entire harvest from their debtors and sell it in more profitable markets or at times of higher prices, including back to producers out of season.[83] Thus lenders had a perverse incentive not to encourage the improvement of cultivation techniques. They could reproduce their income by absorbing the risks of variable profits from production at constant levels of taxation and thus cream off the difference between the marketed surplus and revenue levels over time.[84] Ironically, such increased profitability and improved market access as did come with the commercialisation of agriculture thus served moneylending interests at the direct expense of peasant producers. High land revenue rates, intended to stimulate commodity production, allowed *sowcars* to create a bond with many cultivators by offering to meet the annual demand. When the economic slump of the 1870s started to affect the viability of Ahmednagar peasants and the risks thus became greater, the overextended credit market started to contract too, resulting in the decline in cultivation seen in Chapter 1. The dramatic accentuation of this squeeze when the monsoon failed so widely in 1876 was possibly as much of a trigger of famine as the rain shortage itself, as is argued in Chapter 3.

It was not just lack of autonomy that prevented Ahmednagar *ryots* from taking the risks necessary to take advantage of new market opportunities. In as much as they had a choice, they sought to minimise their risks in the face of the constant possibility of rain failure, rather than to maximise their immediate returns. This does not mean that they failed to obey the basic principles of economic rationality, as prevailing opinions based on classical theory suggested. Revenue Commissioner Havelock grudgingly acknowledged their logic before the famine, musing:

> While the necessity of obtaining a livelihood is the means of keeping people up to the minimum of working power, the hope of enjoying the fruits of one's labours is the means of stimulating, under other favourable circumstances, human exertion to its Maximum. It is not improbable that the uncertainty of the rainfall in the Dekkan, which in one year will give luxuriant crops to light soils and careless tillage, and in a year of drought will refuse anything like an adequate yield to good soil assisted by intelligent and careful husbandry, may be one of the many causes which tend to debase the Maratha peasantry.[85]

The familiar implication that *ryots* may as well be lazy failed to take account of the difficulties of survival on marginal land and the frequency of partial rain failure. Their flexibility in crop selection, including willingness to switch between *jowar* and *bajri* in mid-season, was seen in Chapter 1. In 1882, Elphinston observed that Hindus were naturally conservative in their planting strategies by comparison with 'occidental races',[86] although Martin Ravallion has argued more plausibly that occupation carries more weight than ethnicity, as all poor peasants endeavour to ensure that they are constantly prepared for the possibility of famines.[87] What Havelock came close to admitting was that circumstances were not suitable for speculative planting of more valuable crops. The negative impact of price changes was also recognised by the Bombay Government in the early 1870s. Aware that high prices in the 1860s had been used to justify land revenue increases, they initially considered reducing its incidence now that grain prices were falling.[88]

Plenty of other examples can be found of *ryots'* basic rationality and caution. As seen in Chapter 1, for example, the take-up of over-priced public water resources was minimal except in poor seasons, when the harmful effects of crop failure and, conversely, the extra profits to be made when general yields were low made the premium worth paying. This provoked frustration and contempt from the Bombay Government, but made perfect economic sense.[89] Similarly, it was sensible for peasant households to keep jewellery, although such precautions had limited effectiveness, especially as the government refused relief to any still in possession of ornaments during the famine, when terms of trade for such assets were poor.[90] It may even have been that relations with *sowcars*, seen as exploitative by the British, were another form of insurance in which peasants colluded for as long as creditors could be depended upon to meet the land revenue demand in difficult seasons. This idea is explored further in Chapter 3.

Had they been consulted, the linking of foodgrain markets to global trade and the risks associated with it was the last thing poor peasants in Ahmednagar would have wanted. However, in trying to commercialise the rural economy, the Bombay Government did not particularly want to see the predominance of poor peasants, as is argued in Chapter 4. Yet if poor peasants were forced off the worst land, it was unlikely to be taken up by their wealthier counterparts anyway, resulting in an aggregate increase in the district's vulnerability. Nor could failed peasants make the transition into more competitive sectors than foodgrain production, as local industry was also hit by the global depression.[91] The colonial state was divided on this issue. Those who favoured a progressive agenda of modernisation and commercialisation were content to see the demise of what they saw to be thriftless peasants. Others, including the 1875 Deccan Riots Commissioners, warned against the destruction of cultivators who were chronically poor, but no less shrewd for that, to the advantage of capitalists with no direct interest in agriculture.[92]

When it came to the crux of the matter, then, perspectives on an unsuccessful district like Ahmednagar related to dogma rather than observation. New theories of political economy made *laissez-faire* less a policy, or even a philosophy, than a scientific principle, leaving little room to hear – or observe – those who may not, after all, benefit by it. When agricultural difficulties were reported in the district, they were acknowledged but not associated with the government's strategy.

For Ahmednagar peasants, the global market was presented as an opportunity, but also forced upon them in circumstances which prevented them from taking any advantage from it. Moreover, the chief tool utilised was the land revenue demand, based on prices of commodities they either could not produce, or could not sell at those rates.

Free trade during the famine

Prevailing market-based philosophies were applied more strictly than ever during the famine, when put under the challenge of a potentially expensive crisis situation. As Srinivasa Ambirajan has noted, 'virtually every document relating to the formulation and execution of famine policy over a century referred to the views of Adam Smith and/or John Stuart Mill'.[93] Relief itself was necessarily an intervention yet, just as the state had earlier sought to develop the agricultural sector without investment, now most wanted to rescue it without participation in the grain trade. Private trade was capable of carrying far more than the government possibly could, and it was felt that, if the state interfered, much private trade would be put off. Before the Famine Codes were written, this was not a cut and dried policy, however, and controversy arose when the Government of Madras instantly imported 38,000 tons of grain when the 1876 famine broke out there.[94] This was stopped by John Strachey, who sent Temple to Madras as a special famine envoy to correct their folly, before appointing him as Governor of Bombay the following year. This appointment was itself ironic. In his capacity as Lieutenant-Governor of Bengal, Temple himself had, in 1874, conducted a highly successful but expensive famine campaign in Bihar, which not only imported masses of grain into the scarcity districts, but was left with nearly 100,000 tons of it when the rains returned.[95]

The political storms that resulted from the Bihar and Madras famine campaigns suggest that they were unsuccessful ideological rearguard actions against the new *laissez-faire* orthodoxy and, as such, isolated cases. This ignores, however, the impracticality of complete *laissez-faire* in the chaos of a famine, as a return to Ahmednagar will reveal. As soon as the 1876 monsoon failed, according to the local newspaper *Nagar Samachar*, the Municipal Commissioners there, too, bought 10,000 rupees worth of corn at Nagpur to sell cheaply in the town market. Encouraged by this, a local private merchant similarly undercut going prices by the sale of another 50,000 rupees worth at cost.[96] This was soon stopped, too, but may not have been so stupid as it was deemed, especially when merchants themselves were involved. Prices rise sharply when there is scarcity in a district, and it is well-established that when the commodity in question is food, a combination of panic and hoarding exponentially accelerates that rise. Such rocketing grain prices can be seen during the famine in Ahmednagar in the summary Table 2.7, but extreme high prices, and the causes of them, can also discourage outside traders – or at least diminish the British logic that if prices were high enough, grain supply must surely follow. Grain merchants had to weigh up the potential for profit from famine prices against inherent risks, transaction costs and imperfect information, as well as their other alternatives. While state non-intervention was justified by the argument that famine prices were as consistent as any with the free trade mechanism, Jacomb referred to them as 'extraordinary'; a bureaucratic term commonly used to explain departures from standard policies.[97]

Table 2.7 Ahmednagar famine retail prices (rupees, *annas* and *pice* per standard *maund*)

Foodgrain	1876–77			1877–78		
	Average	Lowest	Highest	Average	Lowest	Highest
Bajri	2-10-2	1-12-2 (August 1876)	3-11-10 (November 1876)	4-2-5	3-2-7 (January 1878)	6-3-10 (September 1877)
Jowar	2-5-0	1-5-3 (May 1876)	3-6-1 (March 1877)	4-2-6	3-2-5 (January 1878)	6-7-3 (September 1877)
Wheat	2-12-1	1-15-6 (April/May 1876)	3-12-1 (November 1876)	4-11-2	3-8-6 (April 1877)	6-1-10 (September 1877)
Gram	2-4-5	1-7-10 (April 1876)	3-5-1 (November 1876)	4-6-4	3-5-8 (April 1877)	5-15-5 (September 1877)

Source: Stewart to Robertson, No. 3195, 22–4 July 1878, Appendix C, p. 361.

First, as Sen's entitlements theory brings out, the chief problem for famine-affected populations is the inability to pay the same excessive prices which were supposed to attract trade.[98] Sellers were inevitably wary of meeting a demand that was not backed by available resources. Second, if the prices had been exaggerated – as the British themselves believed them to be – by hoarders seeking deliberately to manipulate the grain market, then outside traders ran the risk of forcing these stocks out, and thus reducing the famine prices they had come for at a stroke. For an agency which wanted just such a result – such as the state – the same strategy may have made sense, and isolated cases of government grain trading in Ahmednagar did both reduce prices and encourage village *banias* to sell.[99] Third, the costs of transporting grain long distances to remote areas had to be offset against the volatility of famine markets. If the drought ended while food supplies were *en route*, all potential profits could be lost.[100] Furthermore, the globalisation of agricultural markets had already proved a considerable boon to producers and traders in more fertile regions of India, like Punjab. They had no reason to take the significant risk of diverting their high quality wheat (which was anyway not necessarily what the population of Ahmednagar was willing to pay for) to inland districts with which they had never before dealt. They could make a healthy profit, and be sure of future orders, by continuing to supply Europe, or labour colonies like Mauritius. As for the traders in Ahmednagar itself, they naturally had little experience of long-distance dealings, most particularly imports. The markets did not exist. *Jowar* and *bajri* were usually only traded locally, because their bulk meant that transaction costs exceeded their value,[101] even at the highest price levels, as demonstrated during the 1874 Bihar famine. Moreover, many Ahmednagar traders were so small that even they were fearful about their access to new supplies as prices rose and wholesale merchants hoarded their stocks. Thus Jacomb reported that the immediate response to famine prices was for all grain shops to close in panic.[102] Though this was temporary, many small shopkeepers could not buy to sell, while those who still had stocks often withheld them to guarantee their own household needs and to prevent attracting looters, ironically waiting for a fall in prices, to indicate a foreseeable end to the famine crisis.[103]

The problem with the British reliance on the market to solve the famine crisis was that administrators filled with textbook theories based solely on prices had a less sophisticated understanding of the nature of the demand pull than the traders themselves. In an inevitably unequal market, such transaction cost factors as the length and strength of the relationship between buyer and seller, the certainty of payment and the ease and cost of delivery were more important. This had been better recognised before the attempt to create uniform famine principles, during and after the 1876–78 famine. An 1867 report on past famines in Madras by Mr Dalyell pointed out that Adam Smith and J. S. Mill themselves had warned of the exceptions to their general rules on trade, specifically where transport was inadequate, or where fears of famine generated an effectively extraneous distortion of the price mechanism.[104] Similar assertions that the laws of political economy should be relaxed in famine situations were made by Secretaries of State Lord Cranbrook, during the 1866 Orissa famine,[105] and Lord Salisbury after the 1873 crop failure in Bihar.[106] In Bombay in 1876–78, however, private traders, responding to published famine prices, were expected to meet all the demand. Any kind of state interference, it was argued, could only discourage them. This suited a government wanting to give clear and firm orders to its officers, but even supply and demand is not so simple.

Markets depend on confidence, which unlike prices cannot be quantified or published and then expected to deliver results. If the aim is genuinely to facilitate trade, a more complex approach, including occasional intervention, is required. For example, fear played a major part in stifling famine trade. Dealers set prices according to their future expectations, rather than the size of their existing stocks. Thus, the threat of a poor harvest affected prices instantly, although the food on sale was that grown the previous year, or imported. Similarly, prices fell the moment rains returned, even though many cultivators were not in a position to plant new seedgrains, let alone harvest them. There is another complication that belies the simple equation of supply and demand. Some demands have a stronger pull than others and, as with the operation of a fulcrum, equal or lesser weights can prove stronger if they are further from the centre. Even when people are able to pay local famine prices, grain supply may meet demands elsewhere, which are not necessarily reflected in higher prices, but are more powerful. In some cases, this related to political pressure, particularly when the competing demand was from Europe. In others, it was economic logic. Grain sellers had to respect contracts, and also prioritised buyers whose custom was expected to continue year after year. Regular exports continued even from famine districts, as well as from potential suppliers. Bhatia has pointed out that every Famine Commission Report after 1861 noted the contribution of exports to the difficulties faced during famines, although they were never banned.[107] Ironically, British success in persuading the *Nizam* of Hyderabad to adopt a similar free trade stance resulted in 100,000 tons of grain crossing the border into British territory during the distress period, despite famine in his own dominions.[108] Ahmednagar, too, faced an external demand for its produce as the famine wore on. Stewart recorded in 1878 that, although yields were 'not abundant', local traders 'had to supply the wants of other parts of the country where the grain was in greater demand'.[109] Demand, then, was not the same as need. This raises the possibility that famine was exacerbated by dependence on free trade. High prices pulled food in from neighbouring districts until they had too little themselves, thus extending the scarcity, both spatially and temporally, in a series of ripples, spreading out, then bouncing back again and overlapping, as prices went up in areas of supply until they swapped roles with areas of demand. In such a widespread famine as 1876–78, Ahmednagar's place in the pecking order was not clear-cut.

All this is not to say that private trade did not respond to the famine demand in Ahmednagar, especially once relief works provided much of the population with the cash to purchase small quantities of grain. It was just not so smooth or comprehensive as Temple and others claimed. Ahmednagar was 'situated at an inconvenient distance from rail or port',[110] and it was initially feared that local demand would not be met. Though imports did arrive, they had slowed by April 1877, causing prices to rise again.[111] Imports then stopped almost completely after the 1877 harvest, although it was also poor, at the same time that external demand for Ahmednagar grain increased. As a result, cheap foodgrain prices were highest in the district when the famine was believed to be almost over, as seen in Table 2.7. Famine had been said to be characterised by high prices, which generated automatic mitigation through trade. Yet in September 1877, *jowar* and *bajri* prices in Ahmednagar were much higher than they had been the previous year – and even above those for better quality grains – precisely because the district was no longer seen by traders as a priority famine site. Thus grain could only be obtained at arbitrarily spiralling prices which reflected a vicious circle of market failures to supply people's needs. Even boiled down to food supply, then, famine hardship was only indirectly related to the extent of rain failure.

The Government of Bombay also acknowledged 'the failure of the Great Indian Peninsula Railway to meet the demands of the export and famine traffic in foodgrains'.[112] This was partly because of the success of globalised agriculture. So much cotton – including imported piece goods – was being shipped around India by traders with the resources to pay freight rates in advance, that Temple was forced, in September 1877, to impose quotas on commercial trade's use of the railway, in order to ensure that the less competitive – but more urgent – grain supplies could be carried into Bombay Presidency.[113] When justifying this, the Bombay Government demonstrated its chagrin at having to make any intervention in the market: 'It was ... with the utmost reluctance, and only under the strongest sense of the paramount duty of preserving a vast population from starvation, that this Government departed from what experience has shown it to be the most prudent policy.'[114] It did help, but further diminished the diversity of export and through-trade in Ahmednagar, which was supposed to be its best prospect for recovery.[115]

If Temple was embarrassed at this exceptional but public intervention in the market, local officers in Ahmednagar were often inclined to help trade along more directly. In November 1876, Jenkin Jones reported that the grain being supplied to some of his PWD works was not fit for labourers' consumption. He suggested that his staff might do better to buy and sell grain themselves – or even pay workers in kind.[116] The Bombay Government firmly rebuked him for this suggestion, telling him to stop 'meddling with the food supply question'.[117] With due humility, he therefore promised not to interfere again, but sent the government comments he had gathered on the subject from collectors and PWD staff. This included the revelation from Howard, the Executive Engineer of Ahmednagar, that on many of his works 'The grain is Government grain supplied by the Collectors at a fixed rate.'[118] This was later explained to be because 'grain is not procurable in the District in which they are employed'. Far from the market naturally flocking to remote areas where wages were paid, he had to intervene to 'encourage dealers to establish shops within the reach of the work people'.[119] The government responded that they 'forbid officers to take upon themselves the functions of such dealers', and that 'the Executive Engineer of Ahmednuggur should be instructed that he must make arrangements which will obviate the necessity of grain payments on any of his works'.[120]

This would have been difficult without further intervention in the market. The advantage of state grain sales according to Jenkin Jones was that they forced grain prices charged by local *banias* down to the same rates. Having set wages on a sliding scale, allowing relief workers to purchase a fixed amount of grain, the government had laid itself open to exploitation by traders who could demand prices near to works well above those in local markets, without fear of competition. If officers tried to ensure that prices were fair, traders could pull out, leaving no option but state procurement. In the spring of 1877, Jacomb reported that due to 'a thorough combination of the grain dealers a ficticious price had prevailed ... it was found necessary to purchase grain and charge it to the works.'[121] Two months later, Candy warned that all along the Dhond road and railway,

> Bannias now often refuse to sell at all and when they do sell offer 6 seers and 6½ seers for the Rupee. The grain too which is sold is often very bad. I would urge that Government grain be sent ..., the

rate being fixed by you, ... otherwise the coolies run a fair chance of being starved.[122]

On this occasion, with the state also a potential loser because wages would have to increase as prices did, Robertson recommended Candy's suggestion, but only on an *ad hoc* basis.[123] Another two months later, Candy again reported that due to low relief wages, 'Dhond Banias refused to supply grain on the works' causing him to feed workers.[124] Moreover, Assistant Collectors Hamilton and Fforde pointed out that state procurement – and even payments in kind – had been permitted when the relief wage scale was first published, and that it actually saved the state money because they could buy grain cheaper away from the works than labourers could on site.[125] Indeed Jacomb had won approbation for purchasing cheap grain early in the famine to supply the Ahmednagar Relief Fund Committee when prices were high later, earning 1,500 rupees.[126] Thus, while the Madras Government was vilified for importing grain to keep markets open, the Collector of Ahmednagar was quietly permitted to act as a grain trader to make the state a profit but not, later, to guarantee a supply on relief works.

Conclusion

Famine prices were not quite the beacon they were imagined to be for attracting grain into local markets. Rather, they reflected the well-being of those already trading, whose ability to take advantage of naive state policy enabled them to prosper, earning imitation from Jacomb, but condemnation from his colleagues.[127] Just as the Dhond–Manmad railway did not greatly affect Ahmednagar grain trade, the encouragement of private trade during the famine did not enable it to reach all those in need, or even all those concentrated together on state relief works. In both cases, local officers knew this – and sometimes responded accordingly – but could not shake the Bombay Government's confidence in the market. District administrators were advised to toe their superiors' line, while higher officials who dissented from free market logic were dismissed as mavericks. The principle of *laissez-faire* in famines was to be firmly established in the Bombay Famine Code. Yet Temple's subsequent failed attempts to build more railways, and the creation of Agricultural Departments in every province to survey the peasantry, albeit still without financial assistance, suggested that the famine had forced the government to recognise some structural flaws in the agricultural economy. Given the lack of transport facilities in Ahmednagar, the lack of sufficient expenditure – during as well as after the famine – to improve them, and local knowledge that private trade did not meet the needs of famine sufferers adequately, it is hard to disagree with Bhatia's conclusion that the laws of political economy were really used as an excuse: 'It appears that behind the facade of the theoretical argument there was the fear that the Government would have to assume a gigantic financial responsibility in undertaking to feed a vast population during the period of a famine.'[128]

The paradox of Ahmednagar bearing the brunt of an exogenous slump despite its weak integration with wider markets can be explained by three factors. Both producers and consumers were poorly served by transport facilities, few landholders or traders owned significant capital and most *ryots*, as a result, had limited economic autonomy. The price mechanism was not of any benefit to producers. Although at least imports reduced consumers' vulnerability in theory, the nexus of high prices, low yields and fluctuating trade flows during the famine accentuated the possibilities of profit and loss that *ryots* would prefer to avoid. Local prices unaffected by outside trade reduced profit opportunities and could be dangerous – as the famine itself proved. But the impact of exogenous market trends on local producers could be just as harmful. Peasant incomes and security had been devastated by the long 1870s price slump, which far outweighed the indirect advantages of the 1860s boom, which were slow and uneven. On the one hand, as world cotton supplies increased after the end of the American civil war, and demand fell in the global recession, regional demand for cheap foodgrains was also diminished. Income falls were greater than rises – limited by insufficient market integration – had been during the Bombay cotton boom. On the other hand, and more importantly, the tax-induced distress commercialisation of *jowar* and *bajri* production had exacerbated exchange relations in which marketed surpluses fell into the hands of creditors rather than producers, as is seen in the following chapters.

The worst thing about the world market for a district like Ahmednagar, then, was that, being so demand-driven, it failed to respond to the district producers' own needs. It was never really there when it was needed. Whatever Temple claimed, private trade could not really have supplied enough grain to Ahmednagar when there was not even a railway. Even after the famine, grain prices stayed high for a long time. Local producers were devastated economically, no longer able to obtain credit, and out of the habit of cultivation, because it had been pointless to compete – even with limited external trade – when their land was so dry. Before they had recovered anything like their old levels of output, the famine traders had long departed, railway or no railway, while local traders sought to replenish their own stocks before returning to the markets. By 1879, local administrators, who had bemoaned the low prices of the 1870s, recorded slight falls in the prices of *jowar* and *bajri* as 'improvements', demonstrating how the famine had switched attention further away from the interests of cultivators than ever.[129] This has two implications. First that, during and after the famine, *ryots* attracted less concern than landless labourers did and second that distress commercialisation had made a substantial proportion of peasants into net consumers. Eventually, after several poor seasons followed the famine in northern *talukas*, high prices in 1881 attracted imports again, which sent prices tumbling despite low yields, ruining the profits of those who had managed a harvest.[130] With fluctuations in prices reflecting ever more nervous local markets and poor prospects in both good and bad seasons, the *ryots'* situation was as bleak as ever.

Taking Bombay Presidency as a whole and a long time-span, there were areas in which conditions did emerge to allow for a degree of successful market participation. Commercialisation necessarily takes time and indeed may be expected to take longer in a poorly endowed district. It has not been the aim of this chapter, however, to assess the extent to which Ahmednagar was eventually commercialised. The focus has been, rather, on the impact of economic changes on household livelihoods. For particular farmers, the process can be as critical as the outcome. Increases in wealth – or even reductions in vulnerability – between 1860 and 1920 tell us little of what happened to peasants between 1870 and 1884.[131] Later wealth increases, if there were any for the Ahmednagar peasantry, affected later generations. Indeed the implication could well be that poor 1870s cultivators were themselves the obstacle, who could legitimately be sacrificed in the name of progress.

In some ways, Ahmednagar was an especially bad case because the state failed to respond to its particular needs, imposing the same economic rigours upon a district with little capacity to profit as on those that could. But it was not alone in being poor. Its unique features were its virtually undifferentiated peasant poverty and its juxtaposition with successful cotton districts and India's most important port. Thus the contrasting effects of market forces, in a period which included both a boom in global demand for some western Indian agricultural produce and the severe and widespread famine, are brought into relief. This could be analysed in a wider or comparative study, for example by contrasting Ahmednagar with a profitable cotton district like Dharwar, which was caught badly unawares by the famine, and suffered worse mortality than Ahmednagar, but then recovered much more quickly.[132] In considering instead what may be seen as a worst-case scenario of how market forces can be harmful over a period of years, the emphasis rests less on the market than on the state.

Even the most ardent free-trade theorist would not claim that the market could improve the situation for those who cannot make themselves sufficiently competitive to survive in it. The advantage would be for the consumer that uncompetitive producers were driven out. That may make sense on a world scale, but when such global strictures were applied to an almost universally uncompetitive district like Ahmednagar, the nature of their failure was melodramatic indeed. Forcing peasants to rely on external markets was like pushing them towards a cliff, but it did not stop after the famine crisis. Not only did the British somehow believe that they should still be able to fly, they pushed them with the sharp spikes of increased revenue demands and reduced credit opportunities. By adding pressure at the most critical times, these did enough damage by themselves to ensure that small cultivators could never enter the wider marketplace in a position of strength, as is seen in the coming chapters.

3
Rural Moneylending, Credit Legislation and Peasant Protest

Introduction

In previous chapters it has been seen that indebtedness was widespread among Ahmednagar's peasantry and that the agrarian structure of the district was such that rural moneylenders commanded almost all local capital. Though village *sowcars* were mostly small operators, dependent on their urban kinsmen for their own resources, the extent of peasants' thrall allowed monopolistic *Marwari* lenders to control production and extract its surplus value without having a direct interest in cultivation or its improvement. Indeed the moneylenders opposed the state's optimistic hopes for development, which would make their own role less profitable.[1] For centuries, indebtedness has been seen as one of the most serious brakes on rural development in many parts of India and has been associated with peasant poverty and food insecurity. This link was especially strongly made in the case of the 1876–78 famine in the Deccan, with the Famine Commission Report and almost all colonial officers blaming *Marwari* exploitation above other factors for *ryots'* inability to cope with the monsoon failure. The importance attached to this in the report undermines the assumption, particularly by Temple, that peasants – as opposed to landless labourers – suffered relatively little during the famine. Indebtedness was a focus of attention in part because its extent, nature, causes and effects had already been raked over extensively by the state before the famine, as a result of major protests against moneylenders in Ahmednagar and neighbouring Poona district in 1875.

The Deccan riots have been the subject of much historical argument, most of it curiously unrelated to the following year's famine outbreak. Debates have focused on whether lenders' sharp practices or British attempts to regulate them were more to blame for peasants' difficulties.

The idea that legislation had dangerously upset the balance between *sowcars* and peasants was a common theme in the official Deccan Riots Commission Report. The replacement of village *panchayats'* (councils) customary jurisdiction over disputed debts with that of civil courts and the establishment of the right of land transfer so that holdings could be mortgaged for debt led to a considerable degree of self-castigation. To a lesser extent, the *ryotwari* land revenue system was also criticised for encouraging an over-reliance by peasants on exploitative credit. This reflected ambivalence and internal conflict within the colonial state on the issue as well as the commissioners' concern at the extent of debt in the Deccan. Many were willing to condemn usurers as scapegoats for all local problems and themselves for allowing the situation to develop as it had. The report attacked *Marwari* lenders in strong, personal terms: 'His most prominent characteristics are love of gain, and indifference to the opinions and feelings of his neighbour.'[2]

This chapter argues that, while the nature of indebtedness was a problem for many *ryots* – removing any prospect for self-improvement – it was, rather, a sharp decline in the availability of credit that signalled their demise. *Sowcars* had, as Charlesworth contends, played a useful role in providing insurance against the variability of their annual harvests.[3] Moreover, easy and convenient blame of unpopular *Marwaris* let the Government of Bombay off the hook for its own undermining of cultivators via taxation. As well as investigating the causes and nature of rural indebtedness in Ahmednagar and conflicting colonial conceptions of the problem, the state's own attempts to provide credit, and thus replace usurious private capital, are examined. The thinking behind *takavi* loans, both for agricultural improvements and to help cope with crises, is compared with actual practices, showing why most peasants continued to deal with *Marwari* lenders instead. Plans and debates concerning agricultural banks, including specific proposals in Ahmednagar, will also be discussed. These demonstrate, further, the tension between different British ideas on credit, equity and their own role.

The extent and causes of agricultural indebtedness

How critical an issue was rural indebtedness in purely financial terms, in terms of vulnerability and as a constraint on peasant autonomy in economic planning and farming decisions? It is necessary to start by examining the complex nature of rural credit relations and their nexus with other factors like the land revenue. In the words of Revenue Commissioner Havelock, the explanation why *sowcars* treated Ahmednagar *ryots* like 'slaves', unlike their counterparts in more fertile regions like Gujarat, 'does not probably lie easily recognised on the surface, but has to be traced through a network of many varying conditions'.[4] It was estimated that two-thirds of Ahmednagar peasants were indebted in this period.[5] This had been so for a long time, with 'loud and general' bad feeling against *Marwaris* noted in the early 1850s.[6] However, the depth of debt was perceived to be worse by the 1870s, when the average debt was 18 times more than peasants' annual land revenue bill – less than 20 rupees in two-thirds of the cases, which is an indication of how small their holdings were. In Ahmednagar, Collector Boswell showed that mortgage values reached as many as 204 times the annual land revenue demand in Akola *taluka*, and 338 in Sangamner.[7]

The Deccan Riots Commission argued that low prices and poor harvests were serving to 'swell the proportion of embarrassed to solvent ryots',[8] disputing the view that *ryots* who rarely held their sale profits in cash were 'improvident'.[9] This was backed by Assistant Collector Loch in Ahmednagar, who picked the wrong metaphor in praising peasants' 'considerable frugality with a view to providing for a rainy day'.[10] However, it was contradicted by the Famine Commission, which accepted the role of crop failure but saw the other main causes of debts as 'ceremonies' and 'thriftlessness'. Confusing the origins with the processes of debt, the commissioners then discussed the relative importance of 'an oppressive body of middlemen' and 'administrative errors'.[11] They argued that 'too much weight should not be attributed' to rigid land revenue as a cause. If a man spent all his income on himself and then had to borrow to pay his taxes, they argued, the taxes could not be blamed.[12] The implication was that both poverty and debt were individual failures, best dealt with by harsh lessons, thus justifying land revenue rates set to encourage thrift. Not only was this extraordinary in the light of the famine (albeit convenient for the state purse), it reversed the interpretation by the Deccan Riots Commission: 'If it be held that painful experience will teach ... prudence, it must also be shown that the suffering is produced by causes which it would not be possible or right to remove, or that it is justified by the results of its teachings upon the sufferer.'[13] In the view of E. C. Buck of the Government of India, it could not be, because 'a great deal of the improvidence with which the Indian cultivator is charged is simply due to the impossibility of his being provident. How can he look forward when there is no certain prospect before him'?[14]

While some peasants did first borrow for social or ceremonial reasons such as weddings, most debts originated from more mundane needs.

Replacing bullocks or other farming implements was beyond poorer *ryots'* annual budgets. Production shortages could also set off new debt chains. Peasants who had to borrow for seedgrains after the crop failures of 1871–72 were more vulnerable to the next crisis in 1876, when credit was drying up. In some cases, cultivators had no memory of the particular outlay that first got them into debt, in part because, in the absence of bankruptcy laws or even limitation of debts by death until 1859, some debts were immemorial. Other peasants, though, needed to borrow just to pay the land revenue or buy grain for subsistence before the harvest.[15] Large borrowers who retained autonomy and re-invested in their land were a different category,[16] but this was rare in Ahmednagar.

The nature of rural lending

If borrowers were predominantly impoverished, there was a greater variety of categories of moneylenders. Especially after the end of the American civil war, capitalist lending operated through several strata, with *Marwari* village *sowcars* borrowing from their urban kinsmen, or acting as their agents.[17] These credit structures originated from the expansion of cultivation in the Deccan in the 1850s and 1860s which, as Sumit Guha has shown, had been financed by new credit rather than peasant savings.[18] After Sir George Wingate's moderate 1850s revenue survey, combined with rising foodgrain prices, agriculture seemed a plausible prospect for investment even to those with modest accumulated capital. The *Ahmadnagar Gazetteer* records that 'the most unscrupulous class of petty moneylenders increased considerably during the ten years ending 1875'.[19] A more active credit market made it possible for many *ryots* to farm, who could not otherwise have done so, but it also caused two other changes in credit operations. Better legal protection of debts encouraged lenders to deal with people whom they did not know. Moreover, in the absence of agricultural take-off, new ways had to be conceived of ensuring moneylending profits. Interest rates did not come down to reflect the increase in credit supply, because new credit was itself a response to bottomless demand from decreasingly viable cultivators.[20] In sum, the expansion of credit made it a higher risk business for borrowers and lenders alike.

Despite this, Ahmednagar credit markets were uncompetitive. The Deccan *Marwari* community had 'the centre of their exchange and banking business' at Bamburi, in Rahuri *taluka*,[21] and kinship relationships prevented *sowcars* from undercutting each other. As a result, the Deccan

Riots Commission concluded:

> the Marwari, who has almost a monopoly of money-lending in Ahmednagar, is harder and more exacting in his terms than the sowkar of the Poona district, while at the same time the ryot is, if possible, more ignorant and helpless. It is not uncommon for the Ahmednagar ryot to continue paying the assessment of his land after he has transferred the right of occupancy to his mortgagee.[22]

It is, therefore, necessary to question whether moneylending practices were justified in the circumstances of credit being extended to poorer peasants who could not have held land without it. There were three types of interest charges, conforming to the nature of the original loan. *Vyaj* payments were in cash, *manuti* in grain, while most creditors took a combination of the two. Grain payments were usually under the *vadhi didhi* system, whereby borrowing for seeds or food between June and October was repaid in full, plus 50 per cent interest, at that year's harvest. If it failed, cash would be demanded instead, at the ostensibly lower interest rate of 37.5 per cent per annum (half an *anna* in the rupee), which was hard to procure – thus triggering permanent debt. This rate was well above those on existing cash loans, which could be as low as 10 per cent per annum if the peasant's house and possessions were mortgaged, ranging to 33 per cent with no mortgage.[23] In uncertain conditions, though, the *vadhi didhi* rates do not appear excessive. While the term 'usury' was frequent in colonial records, interest levels were not central to critiques of moneylending practice, and British attempts to control them were limited.

Charlesworth's notion that small creditors fulfilled a vital function, for which their high rates of interest rewarded them no more than fairly, needs to be reviewed in this light.[24] He is right that they made no claims to improve agriculture and helped to perpetuate it in difficult conditions. They were not rich, and interest had to reflect risk – though lending was very profitable in the 1860s, with few defaulters, so it could be argued that they should have saved in order to sustain lower profit margins later on. Sympathy for *sowcars* over the terms of the Deccan Agriculturists' Relief Act is rather generous. Though small lenders did lose out, their fate was not so harsh as that of the people depending on them in order to retain their holdings. Moreover, Sumit Guha makes the point that economic analyses of the justice of interest rates overlook the complex ways in which they were agreed by individual lenders and borrowers. In his words, 'It is likely that the neediness of the debtor, his vigilance, the amount of his property, his pugnacity or timidity, and a host of other imponderables determined the actual additions made on account of interest.'[25]

Credit legislation and British self-criticism

Colonial attitudes towards rural moneylending were marked by anti-*Marwari* vitriol. Many administrators had a strong Protestant dislike of indebtedness in general, which, in the influential view of Florence Nightingale, was liable to lead to the 'demoralisation' of lender and borrower as well as economic stagnation.[26] This created a persistent tendency to see credit markets as harmful, despite the belief that every other market was virtuous, and encouraged the British to condemn not only *Marwaris* but also indebted *ryots*. Wingate, for example, speculating on peasants' feelings, wrote:

> He toils that another may rest; he sows that another may reap. Hope leaves him and despair seizes him. The vices of a slave take the place of a freeman's virtues. He feels himself the victim of injustice and tries to revenge himself by cheating his oppressors. As his position cannot be made worse, he grows reckless.[27]

When Ahmednagar Assistant Collector W. F. Sinclair similarly suggested that 'the only ryots at all prosperous are those who, having received some education are able to combat the saukars with their own weapons, fraud, chicanery, and even forgery'; this was not seen as a solution. William Pedder replied on behalf of the Bombay Government that 'it is deplorable to find that moral as well as material degradation has been the result of measures intended to promote the welfare of the people'.[28]

The more pressing concern of the Deccan Riots Commission was that the introduction of civil courts to adjudicate debt claims had unbalanced power relations between lenders and borrowers, by creating a legal basis for debt collection. Enforceable mortgage bonds made it possible for creditors to take more than one and a half times the advance when the season was good. Whereas indebtedness itself long preceded British rule, moneylenders had previously used social means to ensure repayment, with recalcitrant debtors shamed by lenders or hired untouchables sitting in *dharna* (moral fast) at their doors.[29] Thus 'Honesty was the ryot's best policy, and caution was a necessity to the money-lender.'[30] The attempt to regulate debt more systematically was made early under British rule in Bombay Presidency. By giving *sowcars* legal means to realise their loans, Bombay's first Governor Mountstuart Elphinstone had argued that default would become less likely, lending more secure and, ultimately, interest lower.[31] Previously, debt claims were heard by village *panchayats*, which were seen as corruptible because they required gifts before considering cases, and were often beholden to lenders themselves.[32] When civil courts were empowered in 1827, lenders initially reduced their activity for fear of being penalised, but soon realised their advantage under British law. At the same time, *Maratha* lenders – usually village *patels* – who knew their clients personally and could thus use knowledge to defray risk without having to charge extra interest were undermined.[33] An 1843 report into agricultural indebtedness acknowledged that not only was debt more common, but also interest rates higher.[34] Moreover, in Wingate's words, 'The prosperity of the landholder is no longer necessary to the prosperity of the lender. The village lender needs no longer to trust to the landholder's good faith or honesty. Mutual confidence and goodwill have given place to mutual distrust and dislike.'[35]

The Deccan Riots Commission also argued that the over-expansion of credit in the 1860s was due to further ill-judged laws as well as the boom. In 1855, a law restricting interest to 12 per cent was abandoned as unenforceable, giving *Marwaris* free rein. The 1859 Civil Procedure Code and Limitation Act, limiting debt bonds to a maximum of 12 years, only served to undermine *ryots*' security because they could rarely pay off debts within that time. It also encouraged – and sometimes obliged – moneylenders to use courts to close debts. When both parties had an interest in extending the bond, it now had to be re-written. Higher rates of interest were often introduced, and the previous debt could be converted into a new notional loan for a sum far higher than that originally borrowed. Rather than actually calling in mortgages for unpaid debts, *Marwaris* could use the threat of eviction via the courts to hold *ryots* in thrall on their own land, prepared to surrender their entire annual harvest in order to keep their holdings. In the late 1860s, Ahmednagar Collector Norman reported:

> I have often seen the Sowcar sitting in the field while the crop was being reaped which shows that in such cases at least the cultivator is not a free agent, but is compelled to part with his crop at whatever price the Sowcar thinks proper to allow him for it.[36]

Such practices could not be prevented by a new law in 1865 requiring all deeds to be registered, because borrowers were prepared to sign new bonds, either to prevent court action, or in ignorance of their true contents. The commission acknowledged that, in such conditions, lenders had no need to resort to 'serious fraud'.[37] *Ryots* had little knowledge of legal processes, which were intimidating and naturally favoured the educated classes of moneylenders.[38] Sinclair suggested, 'You might as well put a revolver into the hands of an Andaman islander, and tell him to take his remedy on a tiger with it.'[39] Peasants tried to resist the extension of alien civil law from the start,[40] but often failed to contest cases against them, knowing that courts usually accepted fraudulent or extortionate bonds as legal. The colonial conception of fairness on credit matters related to consistent rather than neutral actions by civil courts. In an unequal situation, they therefore consistently favoured one side. Pedder listed the abuses to which *ryots* were routinely subject:

> The passing of a bond by a native of India is often no more value as proof of a debt he thereby acknowledges, than the confession by a man under torture of the crime he is charged with. That the moneylenders do obtain bonds on false pretences; enter in them larger sums than agreed upon; deduct extortionate premiums; give no receipts for payments and then deny them; credit produce at fraudulent prices; retain liquidated bonds and sue upon them; use threats and warrants of imprisonment to extort fresh bonds for sums not advanced; charge interest unstipulated for, over-calculated or in contravention of Hindu Law – these are facts proved by evidence.[41]

The Deccan Riots Commission Report denied that British judges dealt in 'the dry bones of law and procedure, instead of the life-giving meat of equity and justice'.[42] But they approvingly cited a previous Ahmednagar Collector's view that 'it is the law which is at fault in assuming debtor and creditor in this country to be equal, while they are rather in the position of master and slave'.[43]

Such descriptions were common in colonial records throughout the nineteenth century. Only in Ahmednagar was evidence provided to suggest that this went beyond metaphor. *Ryots* and their wives there signed bonds to their *sowcar* for their year's labour, the value of which was only calculated after the harvest, depending on its size. Instead of being paid their wages, the final sum was deducted from their debt accounts.[44] This was a significant difference from deducting the value of the crop, not least because it implied virtual land ownership on the part of the lender. Marx was therefore justified in noting that, while *ryots*' independent holdings meant that they could not formally be subsumed by external capital, 'the moneylender may well extort from him in the form of interest not only his entire surplus labour, but even – to put it in capitalist terms – a part of his wages'.[45] Banaji has convincingly argued on this point, against other Marxist historians, that capitalist exploitation can thus take place where modes of production are not capitalist.[46]

For the British, who wanted to encourage capitalist development, and saw lenders' extraction of surplus value as a key obstacle to it, this was infuriating. Cultivators could not profit from good seasons, or improve their land. More importantly for peasants themselves, the legal right to evict for debt undermined the security for which they had sacrificed their autonomy. While holding *ryots* in thrall gave lenders the best opportunity to profit, they could curtail their involvement in times of hardship – legal bonds being easier to foreclose than social ones. Thus the economic downturn of the mid-1870s greatly increased the likelihood of evictions being carried out. So did anti-*sowcar* legislation. This was most noticeable after the Deccan Agriculturists' Relief Act, but in the year after the Limitation Act, 1860–61, the number of suits for debt in the Deccan also leapt from 91,245 to 312,000. The average remained almost 180,000 a year until the famine.[47]

Land transfers from peasants to moneylenders

The Commission believed that large *Marwaris* started to register land in their own names in Ahmednagar as early as 1855, when restrictions on interest were removed, relegating *Kunbis* to tenants-at-will or labourers.[48] Thousands of *ryots* in neighbouring Thana protested against the risks of land transferability even earlier, in 1840: 'Under the present government, by the sale of our immovable property we are reduced to a starving condition in the same manner, as a tree when its roots are pulled out, dies.'[49] In 1873 the Government of Bombay issued a resolution making it simpler for collectors to record new owners in land registers.[50] This acknowledgement of land transfer is directly at odds with Charlesworth, Morris D., Bernard Cohn and Dharma Kumar's argument that peasant land loss was minimal at this time.[51] Morris's claim that it only occurred during 'unrepresentative' famines is simply wrong in the case of Ahmednagar, though transfers surely increased during the crisis. Cohn's point that power at village levels did not depend on land ownership only makes it more plausible that *Marwaris*, who had few political aspirations, should take possession as their economic might superseded the social influence of *patels*. The process was purely economic, as was the effect on poor *ryots* who had never been empowered.

Land transfer to moneylenders is nonetheless hard to quantify, because it was rarely done through sales. Court decrees of eviction were more often used as threats by lenders than carried out, but not exclusively so. Foreclosures, or claims on 'abandoned' holdings were equally hard to measure. Ownership records were kept only loosely in the 1870s. The Deccan Riots Commission Report pointed out, however, that mortgages should be seen as the final step before transfer, for 'Instances of the redemption of mortgage are almost unknown.'[52] Large transfers to Ahmednagar lenders were reported in the early 1870s. In Parner *taluka*, for example, over a third of land which had been openly transferred was to just six *Marwaris*. The largest, Tularam Karamchand, had acquired at least 659 acres via 55 transactions, and earned 3600 rupees from them in 1875 alone.[53] A third of all recorded transfers in the *taluka* took place in the early 1870s, following a similar surge in the early 1860s.[54]

It is also significant that the framers of the Relief Act were so sanguine about transfers. Charlesworth holds that the Act led to *sowcars* being replaced by big peasant lenders, who foreclosed huge numbers of mortgages during the next famines in western India in the 1890s.[55] This argument rests on colonial comment and statistics from 1896–1902, when the scale of inter-peasant transfers was regretted. Yet, as was seen in Table 1.2, there was already more significant peasant polarisation by 1876 in Satara district, from which Charlesworth draws his evidence, than anywhere else. If transfer figures had existed then, they would probably have been as high as 20 years later. It is possible, therefore, that the variations he detects over time in the Deccan were actually spatial. In Ahmednagar there is little evidence of emerging big peasants at any stage, but the Deccan Riots Commission expressed fears of takeovers by professional lenders.

Marwaris may have disliked cultivation, but they were likely to cut their losses in a crisis – and indeed in anticipation of one – thus exacerbating peasant vulnerability. Nor would the fact that valueless holdings were of little use to *sowcars* have stopped the eviction of at least some *ryots* who failed to pay their dues, *pour encourager les autres*. In Boswell's words, 'As soon as no more can he get out of the orange by additional squeezing, so soon the orange is thrown away.'[56] Collector Jacomb reported in the famine year that 'The Sowkars knew they had nothing to gain by having recourse to the Civil Courts against their impoverished debtors the Kunbis who had no crops and who were compelled in most cases to desert their villages and seek food for themselves and fodder for their cattle elsewhere.'[57] Moreover, smaller lenders running short of capital themselves, had few other assets to sell than holdings mortgaged to them.

Moreover, the effects of the Act that Charlesworth specifically condemns were anticipated even before the riots. In 1874, Havelock suggested that, while limits on mortgage and transfer of land would thwart *sowcars* and improve cultivation, this would be precisely through 'land passing into the hands of wealthier and more intelligent occupants'.[58] There is nothing to suggest that the Government of Bombay feared this prospect in the 1870s, whatever Charlesworth's retrospective opinion. It was consistent with their desire for yeomen peasants to emerge and, more importantly, they recognised that the choice was between land transfers to professional lenders or big peasants. The latter were preferable. Not only were *Marwaris* unpopular, transfers to them like those acknowledged in Ahmednagar before 1875 suggested less acquisitive intent than that the credit system was breaking down. Even *sowcars* told the Deccan Riots Commission that business was becoming impossible because agricultural profits were so low.[59] The riots might, therefore have been directed more against the foreclosure of bonds than their oppressive terms. The fear was less of abuse than ruination.

Land ownership was of the utmost importance to poor farmers. Though surrendering their yields to lenders each year made *ryots* resemble tenants, the commission was only too aware of the likely consequences in difficult years if they became tenants by law: 'if the Deccan ryot is handed over to such landlords as the Marwaris of Parner it may be feared that his fate will be worse than the worst endured by the Irish tenantry of thirty years ago'.[60] The connection with potential famine was, then, understood, and the right of land transfer in the Deccan provoked extended debate within the colonial state. Those who argued that it should be removed, along with imprisonment for debt, believed, in effect, that to be held in thrall was acceptable if it maintained security. According to Assistant Collector Blathwayt, many native officials took this line in Ahmednagar, arguing for land to be exempt from attachment for debt, along with cattle and farming implements.[61] Revenue Commissioner Robertson told the Famine Commission that 'The full freedom of transfer has, in my opinion, been most injurious to Government, and has been the real cause of the present impoverished condition of the Deccan ryot.'[62]

Many others were not so content to sacrifice the opportunity to improve land by allowing poor *ryots* to stay in hock to *sowcars* in perpetuity. They pointed to several land sales recorded at very high prices in Ahmednagar, implying local prosperity – or at least the prospect of it. The natural value of a viable arid holding might be the equivalent of a lifetime of land revenue payments, yet land sales were registered at over a hundred times the annual demand in Jamkhed and Sangamner, and as much as three hundred times more on one occasion in Akola *taluka*.[63] No *ryot* could complain at the loss of his holding if he received a large sum in return. In reality, he got no such thing. While high prices did partially reflect peasant reluctance to sell – as opposed to any objective valuation of their land – they were never actually paid to *ryots*. Seventy per cent of such land sales were further evidence of *Marwari* acquisitions as the final stage of debt processes. Boswell described the build up to land transfers from the start:

> They are in fact mere private compromises by which the ryot, who originally borrowed only 50 rupees of the Marwari; for which moderate sum he mortgaged his field of a rental of 10 rupees a year, but who has found this debt run up in a few years to the preposterous amount of 500 rupees, at last driven by threats of the Civil Court and of being sold out of land, house, cattle, and all he possesses, and induced by false promises that when once the Sowcar has got in name the land which he has long possessed in almost all else, he the ryot will be left in undisturbed occupation of it, consents in an evil hour to transfer to the name of the hated extortioner his hereditary acres. Well may the Sowcar who has purposely, before pushing his claim, run up the debt to an amount which he knew was crushing to the ryot, be satisfied to take instead of his nominally owed 500 rupees land worth perhaps only 200 rupees, and loudly will he boast of his moderation to the ryot.[64]

As Stokes has argued, land transfers were a symptom, as well as a cause, of peasant immiseration. British attempts to treat them as a homogenous problem evaded the issue.[65] In the end, a ban on land transfers was rejected, with even the Riots Commission recognising 'This would be to subvert entirely the form of the property, and such a subversion could hardly be justified by any possible political or economic advantage.'[66] Excepting the Deccan from normal rules, in order to protect vulnerable peasants, proved too great a leap of imagination for the British to agree on it. Even when the Bombay Government sought to amend the Civil Procedure Code solely to prevent the transfer of that part of cultivators' land on which they lived, the Government of India failed to see sufficient grounds for legislation.[67] Thus the argument was lost by those who defended moneylending practice to the extent of preferring stable thrall for peasants to the risk of eviction. By restricting powers of expropriation without annulling the right of land transfer, the Deccan Agriculturists'

Relief Act had a perverse effect. Charlesworth's suggestion that it created a problem where none previously existed carries little weight but it is nonetheless ironic that measures were taken against lenders at a time when peasants themselves wanted more credit, albeit on safer terms.

Transfers unsurprisingly increased during the famine itself. Whereas Morris – and to an extent Charlesworth – see this as a result of crisis, it is possible to view it as part of the cause, given that transfers were already taking place in Ahmednagar before the riots. With the commission's hostile stance adding lenders the incentive to cut and run, the squeeze on both existing debts and further credit made it impossible to override the crop failure. Collector Stewart recorded that while *ryots* were seeking to increase their 'notorious' indebtedness in order to weather the famine 'without relinquishing their occupancies altogether', he had had to fight to prevent legal land sales – or at least to persuade courts to postpone decrees – with limited success.[68]

The Deccan riots

Given the combination of chronic and rapidly emerging problems described in preceding chapters in Ahmednagar, the Deccan riots might be seen as a last cry for help. They were directed specifically at moneylenders but had broader causes and were perhaps intended to be noticed by the Government of Bombay as a warning of their insecurity. The Deccan Riots Commission's evidence regarding land transfers prior to the riots, specifically in Ahmednagar district, is compelling. In these circumstances, rioters could not merely be concerned with the price or availability of grain, as Charlesworth would have it. But nor were they necessarily solely against the various uses of debt bonds, as Ravinder Kumar and David Hardiman suggest.[69] Ian Catanach implicates the government by highlighting shorter-term causes. On the one hand, an edict three months before the riots that land should not be sold by the state for revenue arrears gave *sowcars* an incentive not to meet the demand on peasants' behalf as usual, provoking fury when this led to the seizure of moveable property for non-payment instead. Similarly, the case of a European landholder bankrupted by suspect *Marwari* practice encouraged a rumour – strengthened by the presence of settlement and police officers in the *mofussil*, making enquiries for revenue revisions and the compilation of gazetteers – that the state was planning to legislate against moneylenders, and had ordered the destruction of all bonds.[70]

Boswell, however, saw the rumour as a 'pretext or in some cases ... delusion'.[71] It is argued here that the riots must be seen as the result of long-term fears more than proximate factors. Though peasants' belief that they had the omnipotent state's permission may have made them more willing to protest, this would be at odds with the prior history of peasant insurgency in the presidency. Moreover, peasants' anger that state bailiffs were taking their property suggests that they were also prepared to protest against something to which the state was a party. Indeed, given the Government of Bombay's subsequent acknowledgement that lenders were forced to adopt new strategies because their capital had run out,[72] its own ill-advised action was largely to blame, giving the impression of a combination between state and *sowcar* to strip *ryots'* assets. This was accentuated by Ahmednagar *ryots'* fear of their land revenue being raised as Poona's had been. C. W. Carpenter of the Deccan Riots Commission argued that future hikes were as likely to undermine credit and security as those already implemented.[73]

The Deccan Riots Commission did not see fit to consult peasants on the multiple causes of their poverty. The only debtors' voices they heard were of those jailed for involvement in the riots. While evidence of *Marwari* abuses in their depositions was noted when it supported their agenda, the most common claim of all – each prisoner's complaint that he was innocent but had been convicted on the evidence of lenders who wanted their land – was ignored.[74] The report's policy prescriptions did not address the reasons for chronic indebtedness so much as attack lending itself. To patch a leak from a blocked pipe is to invite an explosion. It is ironic that while the problems of debt and usury have been investigated as contributory factors to the severity of the famine crisis of 1876–78, little has been written about specific links between the causes of the riots and of the famine, from the point of view of either the peasantry or the state. It is interesting to contrast the conclusions of the respective government reports. The Deccan Riots Commission linked usury to the *ryotwari* revenue system, and their combination to the dangerous mood of the petty landed classes.[75] The Famine Commission Report, however, emphasised problems caused by moneylenders specifically in order to refute suggestions that land revenue could be blamed for the general condition of the people. At the same time, it played down connections between chronic poverty and the famine itself.[76] The composition of the two commissions may partially explain different attitudes towards the Bombay revenue system,[77] but it could also be interpreted that, whereas famine was seen as a problem which needed dealing with on its own terms, only insurrection was sufficiently alarming to prompt serious evaluation of state policies with a view to prevention of future recurrences. Even then, the Deccan Riots Commission

Report's moderate proposals for a more flexible revenue demand were not implemented.[78]

Although the riots were quickly put down by force, reinforced by swift trials and harsh punishments, it was known that the problem had not gone away. In Ahmednagar 392 men were arrested, of whom 200 were convicted and jailed although only 22 villages were said to have been fully involved in the riots.[79] Yet, in the Commission's view, 'the warning conveyed by the long catalogue of convictions and punishments and the imposition of punitive police posts, had not extinguished, but only repressed, the violent temper of the cultivators'.[80] Isolated attacks on *Marwaris* continued with, for example, bonds to the value of 9000 rupees being burnt after a single theft in 1878.[81] Nor had the riots come out of the blue. Attacks on moneylenders in Ahmednagar had become increasingly frequent in the early 1870s, with two murdered, one mutilated and one's house set on fire, as well as thefts of bonds and intimidation.[82] The notorious *Bhil dacoit* (bandit) Honya Bhagoji Kenglia also led frequent attacks in hill areas from 1873 until his arrest in 1876.[83] Hardiman suggests that peasants' perceived 'right to subsistence' was threatened by *sowcar* exploitation of British *laissez-faire* policies, with the result that attacks were regular from 1860 onwards.[84] Though the *Ahmadnagar Gazetteer* claimed Honya's gang was only joined by *Kunbis* in 1875,[85] the Deccan riots can nonetheless be seen as a culmination of a burgeoning moral economy which 'involved a critique of capitalist forms of property as well as capitalist systems of law'.[86]

This further suggests that *ryots* were as dissatisfied with their treatment by the state as by *Marwaris*, as does Ravinder Kumar's evidence of emasculated *patels* and *deshmukhs* (village representatives) in Indapur *taluka* (where the riots first started) trying to appeal by petition to the Revenue Department's paternalism, as they had done to the earlier *Maratha* rulers, before taking their own action.[87] It fits, too, with Alex de Waal's somewhat optimistic view that famines were controlled in the twentieth century by the Government of India's need to maintain legitimacy and prevented, after Independence, by a political 'anti-famine contract', empowering vulnerable people to demand intervention successfully in similar circumstances to 1875.[88] This also suggests that E. P. Thompson's conception of moral economy, involving appeals to the squire – or in this case the colonial state – is more useful in the Deccan than James C. Scott's idea of coherent communities inspired by a shared sense of injustice.[89] In the Indian context, peasants wanted not only a moral economy but also *rajadharma*, that is, moral rule.

This does not mean, however, that Ahmednagar *ryots* were ready to rise directly against the imperial state. Local officers were willing to frighten their superiors by referring to land transfers creating 'a revolution in society' and 'a very discontented body' of cultivators,[90] but the Government of Bombay never appeared to worry unduly at the prospect of a general *Kunbi* uprising. Indeed Governor Wodehouse told Secretary of State Cranbrook that the Deccan riots were of no political significance and, but for the Government of India's greater concern, there would not even have been an enquiry.[91] The angriest people were those who had lost their holdings and, with them, their social status and cohesion. Whereas the British yearning for yeomen peasants was unrealistic, there was some truth in the assumption that those retaining land would also respect the law. There was too much to lose, as those jailed after the riots discovered.

The only overt arguments by cultivators against the land revenue demand remained reasonable petitions to unheeding masters. Appeals to British justice against specific *Marwari* actions fell, too, on deaf ears. The response to the riots can only have added to peasants' cumulative sense of having been undermined by the state which was supposed to be helping them. Moneylenders had always been exploitative; the worsening condition of the cultivators was due to something more. The Deccan riots in Ahmednagar were not protests so much as spontaneous expressions of mixed emotions: anger, misery and fear. It is instructive to read Boswell's report that the disturbances 'spread very rapidly, the contagion evidently being highly infectious'.[92] This disease metaphor is particularly interesting in the light of an outbreak of cholera in the district at the same time.[93] Both were to be seen as events that happened to the peasants, without apparent agency, as well as uncontrollable phenomena, bringing out the government's worst neuroses. Nonetheless, Boswell recognised the possibility, under-emphasised by historians, that *ryots* felt themselves to be on the verge of catastrophe, declaring that he 'well knew that nothing but an extreme sense of wrong and a feeling nearly akin to desperation would have goaded them to act as they did'.[94]

The Deccan Agriculturists' Relief Act

It was ironic that the Deccan Riots Commission so strongly criticised the effects of government interventions on credit relations – including many designed to lighten the peasants' burden, such as the moratorium on evictions for revenue arrears – yet concluded that the solution lay in new legislation. Even as a crude measure against exploitation, the Deccan Agriculturists' Relief Act was flawed from the start. As soon as the commission report was published, the Bombay Government's intentions were clear to the *sowcars*, who in many cases sought to cut their losses immediately. Indeed, given that the possibility of a total ban on land transfers was still under discussion, the fear that eviction was the only way to guarantee their investment was even greater than it need have been. Boswell confirmed in 1876 'that the knowledge that some sort of enquiry is being made by Government into these matters has served (as was to be expected) only at present to make the Sowkars more exacting and to render the sufferers more discontented and impatient than ever'.[95] Thus the anticipation of anti-credit measures triggered both extra land transfers to *Marwaris* and a refusal to grant extra credit at the worst possible time, 1875–79. This suggests that a squeeze on existing and new credit exacerbated the severity of the famine, and may have contributed to it.

Anticipating this problem, the Deccan Riots Commission called for immediate interim legislation to prevent evictions.[96] When the legal impossibility of this became obvious, Richey and Pedder proposed compromise deals with lenders, with debts being paid off at four *annas* in the rupee by the state, which would then effectively manage *ryots'* estates under clear rules. Richey argued that 'whatever may be done to put matters on a fairer footing for the future, will be of little avail unless the present embarrassment is somehow relieved'.[97] However, the Bombay Government was more worried by the possibility of becoming directly involved with peasant debts. Hon. A. Rogers claimed that laws controlling *ryots'* expenses would inevitably lead to bans on ceremonial spending, which would be 'the equivalent to social excommunication of every man whose estate came under the proposed act'.[98] While this was a refreshing alternative to blaming peasant 'improvidence', Wodehouse provided the more honest objection that 'if the great mass of the ryots are indebted to the extent now supposed, I think the scheme would involve the Government in responsibilities ... under which it must ultimately break down'. Instead, he argued, in sharp contrast to the commission's conclusions, that 'Our aim must be to maintain friendly relations with the sowkar, to satisfy him that while we will do our best to prevent extortion, we will throw no obstacle in the way of the reasonable investment of his money.'[99]

Such sudden moderation towards the demonised lenders reflected the prioritisation of *laissez-faire* economic strategies over peasant welfare and again suggested an uncomfortable nexus between state and *Marwari* agendas. It did not, however, diminish the punitive prescriptions of the Deccan Agriculturists' Relief Act when it was eventually passed in 1879. Compulsory village registers were created for all debt bonds, while courts were given the right to investigate the history of existing ones and to revise them accordingly, notwithstanding the lack of evidence to contradict *sowcars'* claims. This re-introduced the uncertainty of state intervention in credit markets without regulating them clearly or any more effectively than when they had sought to cap interest rates. Several measures were taken to protect debtors, including the abolition of imprisonment for debt, the extension of the time limit on bonds from three back to 12 years, provisions for insolvency and the requirement for defendants to be present in civil courts. Court fees were also slightly reduced. The combined effect, however, did little to enhance peasant security or to reduce debt, except by discouraging rural credit in the first place. An example of the half-baked logic of the act was the proposal to appoint conciliators to encourage compromises between parties, but also empowered to adjudicate. In addition, village *munsifs* (subordinate civil judges) were to be appointed to hear suits for less than ten rupees, with no professional pleaders allowed. Not only would this ease the chronic backlogs in the civil courts, it was hoped, but they would also recreate the sense of public judgement of the old *panchayats*. Instead, Collector King reported that it was hard to persuade suitable people to take on these tasks.[100] In Akola *taluka*, the entire system had failed by 1881, with conciliators sacked for incompetence and a sub-judge reinstated.[101]

So imperfect were the details of the act that it was amended in 1882, 1886, 1895, 1907 and 1912.[102] Nor did this imply the gradual discovery of a solution to debt. Each new version continued to tinker with the effects rather than the cause. As before, peasants could either struggle by without any financial input, or they could accept exploitative terms. The latter was no worse a choice; for many it was the only one. Douglas Haynes has described how in cotton-growing Khandesh, cultivators who were permanently indebted lost their surplus labour value too, but enjoyed stable relations with paternalistic lenders, who would provide saris or agricultural tools every few years.[103] The difference in Ahmednagar was that peasants were so poor, and their holdings so marginal, that *sowcars* had to squeeze them harder, to the point of threatening their security of tenure, to extract profit. The difference was not of greater cultural tension, or alternative economic logic, but simple poverty. The problems to be solved, therefore, were peasant viability and tenure security. The primary effect of the Deccan Agriculturists' Relief Act was the opposite – to shut off the option of paternalistic exploitation and, with it, all rural credit, instead of creating the conditions for it.

The Deccan Riots Commission was aware that their proposals might reduce the availability of credit, but justified the possibility under similar logic to that which had explained the original institution of civil courts. *Sowcars* would be forced to focus only on more successful peasants, whose credit-worthiness would obviate the need for illegitimate lending practices. Thus only 'bad borrowers' and the 'worst sowcars' would be hurt by the squeeze.[104] Given that the commission's enquiries had paid special attention to Ahmednagar, these false distinctions were extraordinary. Almost all cultivators there were high risk. Sure enough, within a year, King reported that *sowcars* had stopped lending because they feared the act would lessen their chances of recovering either capital or interest. This squeeze was so general as to be 'keenly felt by the great body of the people' and transfers to existing lenders, with holders becoming tenants, had become even more common, leading King to suggest 'if this be the case it would seem as if the Act were likely to precipitate the catastrophe it was created to avert'. In addition 'paralysed' money markets not only reduced stamp fee income but made land revenue harder to collect.[105]

This amounted to an admission of the dependence of many *ryots* on credit to meet their demands, which did not please King's superiors. Robertson wrote in the margin of King's report that the references to widespread credit tightening were contrary to reports he had received from Sholapur and Satara.[106] If true, this would be consistent with the lack of big peasants to take up lending in Ahmednagar. Moreover, as King's examples of land revenue difficulties were in Kopargaon, Sangamner and Rahuri, it suggests a particularly adverse credit market reaction where revenue rises had just been introduced. The Government of Bombay implied that King had judged the act too soon, 'after the first few weeks of uncertainty'.[107] Robertson chastised him on ideological grounds too, declaring that *sowcar* loan terms were 'prejudicial to the cause of improvement' and thus measures against them were fair.[108]

Attacking lenders in this way could not, however, be equated with assisting borrowers, especially in the case of small or marginal creditors, whose profits had already been hit by depression, tax rises and famine. Urban *Marwaris* had cut back their support to rural lenders to reduce their own liabilities and in response to better investment opportunities in Bombay city. At the other end of the scale, impoverished peasants were a worse risk than ever, enhancing the need for stronger guarantees, just when they were being denied. Within two years, Assistant Collector Hamilton reported that 'the race of small money lenders has been utterly crushed.'[109] Though he saw this as a benefit for *ryots*, who had been empowered to refuse interest payments, Collector Elphinston pre-empted Charlesworth by arguing that the collapse of 'a key component of Hindu Society' should rather be seen as a 'calamity'.[110]

Lack of credit prevented recovery from the famine, and sparked off new downward cycles of impoverishment, as *ryots* failed to replace cattle, or even sold them to buy food and seeds and pay the revenue.[111] King added that it also critically decreased the population's capacity to cope with any further economic shock, 'like a famine'.[112] He argued that to reduce such vulnerability, credit had to be rehabilitated 'to the *satisfaction* of creditors', to prevent them from withdrawing or pushing holders out.[113] This at once jarred with Robertson and matched the earlier argument of Wodehouse. It is possible to see a group as contributors to famine and still argue, as David Keen has convincingly done, that the best way to prevent famine is to initiate policies in their favour, thus obviating their need to exploit potential famine victims so harshly.[114] But British confusion over rural credit went further than that. They were not even sure if it did cause, or palliate, famine vulnerability.

The nexus of colonial revenue with peasant indebtedness

King's view that creditors were essential to the smooth collection of the land revenue may not have been popular with the Government of Bombay, but the Deccan Riots Commission had gone further in linking the questions of debt and taxation. Some of peasants' problems were perennial, but there were also several causes 'associated more or less with our laws or administration'.[115] The revenue system as well as debt legislation had tied people into broader credit markets, bringing the downturn of the 1870s to districts that may not have been affected if lending was still done at village level to known borrowers. Because land revenue was set at fixed amounts and times, it inevitably necessitated debt in poor or late harvests, as few had the resources to store grain after good seasons. The report nicely condemned the Survey Department's logic in setting rates designed to be fair on average, by recalling the local fable of a man who drowned trying to ford a river, after calculating that his height was greater than the average depth of the water.[116] The rigidity of the revenue system is explored further in Chapter 4.

Commentaries on receipts from court fees also reveal colonial ambivalence and self-criticism. When the figures increased, the implication was that more *ryots* were being sued for debt and either losing property or being pressured into signing increased bonds. The 16 per cent rise in Ahmednagar court fees before the Deccan Agriculturists' Relief Act was passed, for example, was said to be 'unfortunately due to increased legislation and the increased number of applications for execution of decrees'.[117] When court fees went down, however, as they did both before the famine crisis and after the passage of the act, it signalled the withdrawal of credit as much as reduced pressure for collection. In 1875, Pedder suggested that, far from reflecting an end to *sowcar* litigiousness and *ryot* improvidence, 'the decline in the number and value of money bonds only shows that the credit of the peasantry is becoming exhausted, and that suits to recover money are fewer, only because many of the people are no longer worth suing'.[118]

British confusion was in part because these figures reflected both new loans and efforts to recover them and also because they were unwilling to accept the value of *sowcars'* role. King revealed another concern in 1880 in praising the Relief Act for checking debt litigation but blaming it for low court receipts.[119] The Bombay Government remained as preoccupied as ever with its overall income, making large court receipts a benefit – an irony given that *panchayats'* jurisdiction had been questioned precisely because 'bribes' were demanded in advance. Court costs were often condemned at district level. Boswell opined that 'The excessive costs of litigation appear to me to deserve to stand first among the causes of the failure of our civil justice.'[120] Because it had a financial interest, no matter how small, the state was not able to take the objective stance it proclaimed on this complex issue. An appendix to the Famine Commission Report noted that the High (Appellate) and *Mofussil* Courts made a combined annual profit of 460,000 rupees in Bombay Presidency, from 'the most necessitous class in the country'.[121] Boswell noted that the population of Bhatodi and Athwad villages, recently transferred into Ahmednagar District from the *Nizam's* Dominions, feared the new prospect of British civil jurisdiction more than *sowcars* themselves.[122]

The Bombay Government sought to portray any apparent legal bias towards lenders as accidental, because wily *Marwaris* were exploiting loopholes that were hard to close. But it was not coincidence that they should be beneficiaries of British colonialism. Capitalists seeking to gain from agricultural production, but not to protect it in the longer term, were in a similar position, both ideologically and in practice, to the state itself. Lenders gained by taking advantage of economic opportunities that had been deliberately created by the rule of law and a market-based developmental agenda. Although extracting agrarian profits without re-investing was necessarily anti-developmental, the state was doing much the same thing. The British did not attack *Marwaris* to protect Ahmednagar cultivators, so much as to assert their own rights to peasants' limited surpluses. The greatest tension was between the state's own economic logic and anti-*Marwari* rhetoric, which was strongly felt but superficial. A deeper critique of proto-capitalist practice would have directly undermined the imperial agenda too, hence the dismissal of the Deccan Riots Commission's uneasy ventures into criticism of *ryotwari* land revenue policy. Thus, while Charlesworth's sympathy for the proliferation of new, small lenders remains rose-tinted, his argument that they were not the ones to blame for peasant hardship is fair. They reacted to conditions of stagnation rather than creating them. Keeping peasants in thrall was an economically logical way to utilise capital in the high risk environment of Deccan agriculture, which also assisted some *ryots* in retaining their poor holdings. It was attractive because the Bombay Government legislation – as opposed to rhetoric – supported moneylenders in practice.[123] Mortgages had no value in *zemindari* provinces (where all cultivators were tenants) and would not have had in Bombay Presidency if land transfer rights had not been conferred on the peasantry.

The relationship between moneylenders and the state was thus a complicated one featuring, at different times, condemnation, co-operation and competition. David Washbrook has argued that British colonialism in India depended on the success of its partnerships with existing or emerging political and economic elites, of which the *Marwaris* were one.[124] They may have been an atypical example, at least in the Deccan, which was marked by the weakness of its own pre-colonial elites, but they were not unique among British partners in inspiring more hostility from the government than trust. Middlemen were often essential, but always in the way of both the idea of uninterrupted markets and the direct flow of imperial gains. It was frustration with *zemindari* middlemen (landlords paying fixed tax) that led to the *ryotwari* revenue system, and the realisation that a middleman class had arisen under it was horrifying to many Bombay administrators. What this reflected, though, was not *Marwaris'* capacity to undermine the state's desire for endogenous growth but its impossibility in the first place. In the absence of any realistic prospect of capitalist farming in Ahmednagar's inhospitable terrain, both rulers and lenders had facilitated smallholder cultivation in order to get a meagre return, and collaborated uneasily in that task so long as the margin remained.

Colonial attacks on moneylenders often came in specific contexts where their actions undermined revenue collection. In 1873, a *sowcar* was convicted of theft, imprisoned for a month and fined 30 rupees after he took a debtors' crop that had already been attached for sale for revenue arrears, the Government of Bombay insisting that they always had the prior claim.[125] The *sowcar* whose attempt to attach a European lady's gown and a gentleman's hat allegedly provoked the Deccan riots, was labelled 'insolent'.[126] The fluctuations from co-operation to competition between state and lenders operated like a pincer, in which the Ahmednagar peasantry was caught. Depending on the individual *sowcar* and state agent concerned, this could amount to a stabilising structure, or the means by which *ryots* were crushed. If, after a bad season, the state revenue demand constituted a peasant's entire surplus, its payment depended on the lender's goodwill. If he saw no long-term prospect, or needed a short-term return himself, he could refuse to assist, or extract additionally, first from the peasant's own subsistence needs, then from his productive assets. This happened on a wide scale, for varying and cumulative reasons, every year from 1875 to 1880.

State credit schemes

The extent to which the government missed the point and failed to act effectively was highlighted by the attempt to provide rural credit itself. Loans could be given to peasants by the government in the form of *takavi*. Although it was hoped, especially in the credit crises of the 1870s, that these might obviate the need to turn to *Marwari* exploiters, they were not usually granted for the same primary purpose – to help poor cultivators through hunger months and years. The main aim of *takavi* loans was to encourage direct *ryot* investment in their holdings, in order to improve agricultural productivity and thus peasant accumulation. Further, their allocation was limited, both by the state's resources and its prescriptions, although those who took loans for wells were exempt from having their land re-classified by the Survey Department as *bagayet*, irrigated, which attracted much higher rates of taxation. That *takavi* was used to carry out colonial agendas for *ryots*, rather than in response to their own needs, was confirmed by the alternative use of loans to sedentise tribal areas and to enable native public servants to build houses.[127]

Comparisons with *sowcar* credit were thus problematic from the start, but it was specifically intended that the 1871 Land Improvement Act, by its clear and fair processes for both granting and recovering loans, would make *takavi* the more attractive option for peasants. Whereas *Marwaris* fixed interest arbitrarily, leaving themselves room for exploitation, government loans were fixed at the low rate of 6.25 per cent. *Takavi* loans had been common in pre-colonial India, and entrenched in British law in Bombay in 1827, but Act XXVI of 1871 was written by John Strachey – a key figure in the creation of the Famine Codes, for similar reasons – to provide clarity and imperial uniformity.[128] The Government of India was influenced more by the 1864 Land Improvement Act in England than by administrators in Bombay, who found the new legislation limiting in its criteria for grants of loans. Discretion over interest rates was removed,[129] totals for new wells were capped and loans for exceptional purposes disallowed. For example, sanction was denied for a grant to Krishnarao Ram, an Ahmednagar *jagirdar* (holder of an assignment of land revenue), to distribute to the cultivators of his villages.[130] The Bombay Government Chief Secretary E. W. Ravenscroft reminded the Government of India that loans were not a 'permanent burden on the imperial finances'.[131]

Unsurprisingly, loans under the Land Improvement Act proved unpopular. The Famine Commission reported that so little had been granted by 1880 as to 'bear no proportion whatever to the need which the country has of capital to carry out material improvements'.[132] When provincial governments were asked for explanations of the persistent low take-up of *takavi*, Bombay offered three: the low quality of land – making investment futile – existing indebtedness and 'the natural apathy of the Asiatics'. Race was perceived to be such a relevant factor that native subordinates should not be used to advertise the scheme, for 'The native farmer will seldom be induced to undertake costly improvements unless he is stimulated by the advice and assistance of the Assistant Collector or a special officer.'[133] This was almost the reverse of the truth in Ahmednagar. Elphinston ascribed the unusually high volume of loans in Newasa *taluka* to the exertions of the *mamlatdar*.[134] British officers at various levels failed over time to advertise the availability of *takavi*, discouraged applications and rejected some that were made. The peasants themselves preferred to deal with native lenders anyway, even if only as the devils they knew. Colonel Anderson, the Survey Commissioner who was in the process of increasing Ahmednagar revenue demands beyond reasonable logic, observed without apparent irony that 'people recoil from the idea of involving themselves in money transactions with Government, which is no more than what I should expect'.[135] While they could scarcely believe that crooked *sowcars* served *ryots* better, it suited the Government of Bombay and its agents not to extend themselves too far into the credit market, when there was too much other work to do and resources were restricted from above.

There was, however, a proclaimed desire to operate the scheme beyond the level of initial peasant demand. So few suitable applications for loans were received throughout the presidency in the early 1870s that exceptions were made to support *zemindars* in Karachi and Sind, whose crops, houses and canals had been damaged by drought, rats and floods.[136] The Government of India regarded these as acceptable because they involved little risk, although they had been reluctant to sanction loans for relief purposes to poorer holders. As Mr Trevor, Assistant Collector of Nasik, Ahmednagar's northern neighbour, complained, *takavi* loans generally were

> better adapted to meet the case of comparatively large works undertaken by men of enterprise and intelligence than to encourage poor and ignorant cultivators to have recourse to Government aid for petty improvements they may be inclined to undertake, and it is improvements of this latter class, individually trifling, but of enormous importance if they can be generally extended, which it is specifically desirable to promote.[137]

His Collector, H. Erskine, argued that at least these grants to *zemindars* should mean the principle of *takavi* for drought or flood, though not stated in the act, had been established for all. Distress was particularly liable to throw *ryots* into *sowcar* hands, but could also provide an opportunity to popularise alternative government loans in the longer term.[138] Though Erskine was allowed to loan up to 20,000 rupees that year, Revenue Commissioner Rogers attacked such assumptions regarding the ideal targets for *takavi*: 'I think [Erskine's] assistants appear a little too anxious to press the advance upon the Ryots – If the Act is not very judiciously worked, it will encourage a class of pauper cultivators whose proper place is in the labor market.'[139]

This particularly unfortunate expression of the Government of Bombay's unwillingness to support poor *ryots* in this period showed why *takavi* loans were never likely to be extensively granted. Moreover, peasants showed little desire to re-invest in their lands with the help of cheaper credit. This was not surprising in a period of depression, but there were several additional factors that made *takavi* loans unattractive despite the low interest on them. They had to be repaid within three years, which left little leeway if constructing a well took time, or profits failed to increase immediately for external reasons, such as falling prices. As repayments had to come from improved surpluses, loans could not be taken to tide *ryots* over hard seasons – or indeed for irrigation works designed primarily for risk prevention. The short time frame also meant that repayment instalments were relatively large and started very soon after the loan had been taken. Although the government could suspend them in bad years, like the land revenue, the period of each loan was strictly enforced. As King noted, this was a key reason why most *ryots* chose to deal with the 'rapacity' of *Marwaris* who would 'give them a long day', even when this meant permanent indebtedness, rather than a government which, no matter how fair, was 'inexorable in demanding prompt payment'.[140]

The high risks associated with rigid terms of recovery were accentuated by equally strict rules concerning collateral. The Government of India emphasised that the land to be improved should itself be 'absolutely hypothecated' to the state, in order to enshrine the right to evict non-payers.[141] Additionally, the Bombay Government's rules under the Act declared that 'No advance shall be made unless the value of the security accepted exceeds by at least one-fourth the amount of the advance.'[142] Once again, this made it far harder for poorer *ryots* to take loans. To cover themselves against the combined threat of fixed repayment terms and eviction for failing to meet them, those cultivators who did apply for state credit frequently asked *sowcars* to underwrite them. This defeated the stated aim of *takavi* to undermine private lenders, who could then take the peasants' crops in return as if they had loaned the money themselves. After the passage of the Deccan Agriculturists' Relief Act, the Government of Bombay issued a resolution banning *sowcars* from standing security, only to be surprised when King reported that it resulted in lower demand for *takavi* loans in Ahmednagar than ever.[143]

Even cultivators sufficiently established to benefit by loans for land improvement without risking everything often still turned to *Marwaris* ahead of *takavi* loans. So fearful were the British of default that they were not content to rely on strict rules. Before any loan was considered, 'a minute and troublesome inquiry' had to be made into 'the nature of the applicant's tenure and its value',[144] 'the correctness of the facts alleged in the application' and his own background and creditworthiness.[145] If the loan requested was for over 500 rupees, this investigation involved public notices and consultations, making the application process intrusive, over-complicated and painfully slow. Holdings were also inspected after loans had been given, to ensure that adequate progress was being made on the prescribed works. Notwithstanding the British penchants for excessive bureaucracy and mistrusting cultivators, this belt and braces approach was almost designed to limit grants of *takavi*. If applicants were to be checked for their solvency, there was no need to fix repayment dates, or even sums. Like *sowcars*, the government might reasonably have claimed a share of future profits – up to a set limit – instead of fixed repayments, ensuring that they did not overstretch borrowers. This was not implausible. A suggestion from the Government of Madras, for example, that it could be preferable to ask no repayment at all, but charge full wet land revenue rates as soon as the well was complete, won sympathy but not backing within the Government of India.[146]

This would still have done little to assist poor *ryots*. A loan system predicated on the certainty of recovery – far more than private credit was – could never help them. Poor, arid areas, like much of Ahmednagar, are by definition bad risks. Any investment to develop an impoverished district, or its inhabitants, must accommodate the risk of non-repayment or fail itself. The only effective credit in Ahmednagar would have been generous credit. *Takavi* was as mean in principle as the worst *Marwari* bond, and much less readily given. After four years of exhortations to collectors to grant more loans, resulting in requests for 1875–76 totalling 16,000 rupees from Boswell, and 115,500 rupees throughout the presidency, the Accountant General of Bombay told his Revenue Department, 'There are no funds available for these advances.'[147]

For a variety of reasons then, *takavi* loans could neither assist the advancement of Ahmednagar *ryots* during the early 1870s nor reduce indebtedness, which was seen as a crucial aspect of peasant famine vulnerability. The desire for wells did increase during the famine crisis, however, and total improvement loans in the district leapt from 5,690 rupees in 1875–76 to 20,025 rupees in 1876–77.[148] Robertson claimed this as a breakthrough, patronisingly asserting, 'if the famine has only taught the cultivators to take advantage of the water so easily obtainable by a little digging, it will have conferred no small benefit on the Country generally'.[149] Yet loans to peasants intent on reducing risk, rather than increasing profits, had never been previously provided. This was explained as late as 1882 by C. P. Ilbert, Legislative Member of the Governor-General's Council, in terms of the state's dual role when making agricultural loans – as capitalist lender and as part owner of the soil, with an interest in its improvement. In the latter role, unprofitable loans were justified, but not in the former. The 1871 Act, he explained, saw the state solely in the former, and it would unnecessarily 'mix up loans and revenue' to do otherwise.[150]

Michelle McAlpin suggests that lessons were nonetheless learned from the 1880 Famine Commission Report.[151] Under Act XV of 1880, later confirmed by the 1884 Agriculturists' Loan Act, loans were given for fodder, seedgrains, tools and new cattle to aid cultivation during famine crises, and accelerate re-growth after them.[152] This was not a new idea. Such loans were excluded from the 1871 Land Improvement Act, but were central to pre-colonial conceptions of *takavi* and, more pertinently, were the subject of a dispute between the governments of

India and Bombay during the 1871 scarcity. A proposal from Ashburner, when Collector of Khandesh, for loans to *patels* who were starving on their holdings rather than accepting relief was approved in Bombay, but condemned furiously from above as 'insufficient to justify a deviation from the true principles on which advances to agriculturists should be made'.[153] Ashburner replied, with exasperated foresight, 'The question raised by the Government of India whether there was or was not urgent necessity for the relief afforded, will depend on the degree of starvation and distress to which the country should be reduced before Government should interfere.'

This argument continued, and in 1875 Pedder referred the central government to Viceroy Mayo's speech on the 1871 Act, in which he specifically denied that the intention was to end the tradition of loans 'that may not strictly come under the head of agricultural improvements ... particularly under pressure of famine and distress'.[154] The Famine Commission Report added that loans to landed classes should not only be 'part of the regular system of famine relief' but also 'liberally extended and prolonged till the effects of the famine have passed away'.[155] The Government of India still disputed both of these statements when the 1884 Bill was under consideration. Holderness of the central Revenue and Agriculture Department insisted that loans for improvements and for purchases should be firmly distinguished because 'The Famine Commission deprecate systematic loans of the latter class, as tending to demoralise the people, and the same idea was expressed by John Strachey, the author of the Land Improvement Act, when introducing the Bill in Lord Mayo's Council.'[156] Under the final 1884 Act, moreover, loans were still restricted to those able to convince administrators of their ability to repay.

This says much about the dangers of taking the Famine Commission Report at face value, but less about grants of relief loans in the *mofussil* during the 1876–78 famine crisis. Loans to the extent of 20,000 rupees were in fact allotted from Bombay to Ahmednagar for 1876–77 for fodder, but they were recorded under a separate head from 'ordinary tacavi'.[157] This was impressive, although initially its effects were limited. Hamilton reported that he had granted 6,954 rupees, in sums from 5 to 50, to just 203 farmers in his four *talukas*, some for 'ordinary repairs' to wells instead of fodder. Moreover, under the insecure circumstances, he had taken bonds for repayment within two years instead of three in all cases.[158] Woodburn, his counterpart in the worse hit southern *talukas* of Shrigonda, Karjat and Jamkhed, granted only 4,700 rupees for fodder, almost entirely to better-off *ryots* whose garden crops had not been destroyed.[159] Subsequently, a further 29,500 rupees were allotted to these *talukas* for seeds and new bullocks, which were 'much needed' and widely taken after the 1877 rains, undoubtedly helping *ryots* to renew cultivation.[160]

The effectiveness of loans for seeds was not accepted more generally. Indeed, it is possible that they were limited to Ahmednagar. Though McAlpin mentions them in western India in 1876–78, Bhatia asserts that they were only given in Madras Presidency: 'No use seems to have been made of this method of relief in the Bombay Presidency during the same famine.'[161] Furthermore, Temple told the Famine Commission that peasants would have abused the right to loans they did not need to survive. When it was suggested that the strategy might allow cultivation to start again sooner, he emphasised the Government of Bombay's limited resources.[162] This new attitude at provincial level is significant. After years of debate over whether relief *takavi* was fiscally justifiable, the famine turned the consensus among senior Bombay administrators against it during the crisis, after Temple's arrival. Given the predominance of stringent attitudes at both imperial and provincial levels, it is thus hard to imagine new enlightened approaches to *takavi* being entrenched shortly afterwards in new acts.

Nonetheless, Ahmednagar officers continued to grant seed and bullock loans each year after the famine – though still permissible under Bombay rules only in 'exceptional distress and scarcity'[163] – in larger amounts than those for land improvements. This can be seen in Table 3.1.

This was in part the result of an 1877 resolution that *takavi* loans under the Land Improvement Act should themselves be 'confined to

Table 3.1 Total government loans (rupees), Ahmednagar district, 1877–82

Year	Land improvement (*takavi*)	Loans for seeds, cattle etc.
1877–78	700	7,303
1878–79	1,350	3,782
1879–80	2,250	6,540
1880–81	1,300	4,630
1881–82	5,705	2,515

Sources: King to Robertson, No. 3140, 19–23 July 1879, p. 20; King to Robertson, No. 4584, 22–5 July 1881, p. 55; Elphinston to Robertson, No. 5730, 20 July 1882, p. 75.

special and exceptional cases'.[164] Additionally, there was great demand in the long term for seed and bullock loans. Assistant Collector Candy typically reported in 1879 that 'The sums at my disposal have not been equal to the demands as many ryots are in want of bullocks in consequence of losses during the famine.'[165] Even in 1882, when grants fell, Elphinston blamed *mamlatdars* for being too slow to check that loans would be recoverable.[166] Robertson replied that such caution was praiseworthy and 'necessary to save the State from eventual loss'.[167] This suggests that the rules remained just as strict as before the famine, in continuing conflict with the needs of *ryots*. For all the unattractive aspects of *takavi* for smallholders, in this period it was the government who kept the figures down on the grounds of risk, rather than low demand. Annual reports do not show the numbers of applicants for loans – and therefore how many were turned down. They do reveal, however, that larger amounts were recovered from old loans than were lent, in every year from the famine until 1881–82, when loans exceeded recoveries only in the northern *talukas* where land revenue rates had recently been revised. Thus, although state loans for reconstruction in Ahmednagar were atypically generous, they were scarcely enormous in the context of increased demand for state credit as the Deccan Agriculturists' Relief Act was passed. There was greater concern, after the expensive famine, for the limits of the government's own coffers.

Loans, both for seeds and wells, were thus contested and often arbitrary, being given more according to the predilections of the assistant collectors responsible than any clear policy. In 1879, for example, King noted that far more loans were granted under both heads in northern *talukas* than those that had been affected more by the famine. The reason could only be a 'difference in the principle on which applications are met in the two divisions'.[168] In other words, while Candy was moderately liberal, his counterpart Hamilton tended to loan as little as possible. King stated his preference for Candy's approach, suggesting that 'If he has been careful in the matter of security, his method deserves imitation,'[169] whereas Hamilton's application of rules was perhaps 'not suitable to the condition of the people'.[170] While the rules of *takavi* remained restrictive, Candy was not criticised by his Collector the following year for targeting loans at those 'whose credit is small in the market and who find difficulty in getting loans from Sowkars', adding 'I do not allow these advances to be made if possible to well to do persons.'[171]

This suggests administrative ambivalence, especially between different rungs of the colonial hierarchy, but not necessarily the attitude shift necessary to make *takavi* the effective tool which McAlpin believes it became after the famine. The Famine Commission Report did argue for a radical overhaul: 'While all needful precautions are taken to secure the State from loss, every unnecessary impediment should be removed which now makes the people unwilling to apply for such advances.'[172] In addition, Holderness suggested that the renewed rain failures of 1881–82 in Ahmednagar and surrounding districts, 'may have made the Bombay Government anxious to make advances'.[173] The 1883 Land Improvements Loan Act did also remove some obstacles to the taking of loans, as well as granting more autonomy to provincial governments, if not to district officers. The maximum length of repayment periods was extended to 35 years and interest, while still to be set generally at 6.25 per cent, could be lowered or waived in special cases. In addition, the principle of not recategorising land irrigated as a result of new works for revenue purposes was entrenched.[174] As Secretary of State Kimberley pointed out, however, this latter point should not be an aspect of *takavi* legislation, as it ought to apply to all wells built by *ryots*, however they were financed.[175]

All of this was useful, but did not overcome the basic problem that if loans were only given to peasants whose solvency was certain, they could only be given to the better off, and then after lengthy and intrusive enquiries. The problems of famine and poverty in the Deccan were in the Government of India's mind, but no leap of faith – to subsidise poor *ryots* as Candy wanted to, or to take the risk of default on board – was made by the upgraded legislation. The Finance Member of the Government of India, Major Baring, confessed in 1882, 'I have so little confidence in the system of Government advances ever proving very successful, that I am much inclined to doubt whether it is worth while going to the trouble of altering the existing law.'[176] Moreover, when the threat of state losses and *ryots'* need came into conflict, the priority was unchanged. Even as the Famine Commission Report called for the removal of obstacles to loans, it warned against the 'dangers of excess' in giving them:

> There could be no greater encouragement to unthrift and recklessness among the agriculturists than the knowledge that they have no need to accumulate capital to meet any misfortune that may befall them, but that they can always rely on obtaining from the Government the money they require on comparatively easy terms.[177]

No matter that the view in Ahmednagar during the famine was that peasants refused free seedgrains out of 'honest pride', when they would have accepted loans to buy them.[178]

To be fair, McAlpin notes a significant increase in *takavi* loans in Bombay Presidency only after 1890–91, perhaps because villagers had worked out by then how to apply for it.[179] While this explanation is unconvincing, and the precise date does not sit easily with the failure of the Bombay Government to control the famines of 1896–98 and 1899–1901 in the presidency, it is indeed probable that the *takavi* loan system became more effective some time after 1884. This does not mean, however, that British rural credit policy was always along the right lines, or made effective by the acts of the early 1880s, as McAlpin infers. Rather than an effective *takavi* system gradually bearing fruit over decades, it is more likely that specific changes to it as late as the turn of the century finally made it an effective famine prophylactic. The failure to use *takavi* adequately to ease poor *ryots'* plight before, during or after the 1876–78 famine crisis was to all intents and purposes deliberate. The fact that some seed loans were given in Ahmednagar between 1877 and 1884 suggests a degree of contrary agency on the part of local officers, though its scale and effectiveness was constrained. It might therefore be suggested that when *takavi* did become more popular and successful it was because district officials became able to administer it with fewer imposed limits.

A proposed agricultural bank in Ahmednagar

Long-running debates over alternative forms of state credit also reached a peak at this time. In 1863, a proposal for regulated pawnbroking, with profits going to municipalities, had won support from then Collectors Ashburner and Robertson, but was rejected as liable to favour *sowcars*.[180] Agricultural banks were proposed in 1858 by Jacomb, when he was the Third Assistant Collector of Ahmednagar, but blocked on the grounds that 'it is not likely that the Home Government would countenance any scheme which, although deserving encouragement on its merits, ought, according to acknowledged principles, to be left entirely to private enterprise'.[181] This idea was revived in an appendix to the Famine Commission Report by Syud Ahmed Khan Bahadur, suggesting state regulation to encourage existing moneylenders to pool resources and thus profit without charging such high interest.[182] Shortly afterwards Ahmednagar Judge Sir William Wedderburn won the Government of India's approval in principle for a more specific proposal for an experimental bank in Nagar *taluka*, which would take on all existing debts and be contractually committed to fair practice.[183] Ahmednagar was seen as appropriate precisely because of its poverty and history of both famine and riots, though any hope that the new bank might resolve the chronic problem of agrarian indebtedness was set against a commitment to be profitable.[184] Wedderburn succeeded in recruiting support for his idea from John Bright MP, Famine Commissioner Sir James Caird, *The Times*, *Daily News* and *Standard* newspapers, but most importantly from Sir Nathaniel de Rothschild MP, who told Wedderburn that raising capital would be no problem under his terms.[185] The Government of Bombay was less convinced, declaring 'A district in which a large proportion of the people have no thought beyond the immediate future can scarcely be considered a favourable field for banks of any kind, and if they do not succeed when conducted on purely commercial principles, they are not likely to succeed at all.'[186]

That both sides should use widespread privation in Ahmednagar in support of their case is instructive given the general tendency to exaggerate the district's recovery since the famine crisis. It also highlights the way in which objections to the idea of an agricultural bank there were intertwined with the very reasons for the proposal. If a bank was necessary it was because *sowcars* were harming rather than helping peasants, and state loans had not provided a viable alternative. Yet the solution was to impose a similar partnership of private capital and public management, both of which provoked considerable unease. Despite Rothschild's patronage, capital had to be raised locally, from *Marwari* lenders or members of the *Poona Sarvajanik Sabha*, who would then have to be underwritten by the state, at least to the extent of assisting with the collection of debts. The danger of allying too closely with *sowcars* was obvious to many, including a member of another banking family, Baring: 'Government, by becoming the agent of the banks, incurs all the odium, which in India, even to a greater extent than elsewhere, falls on the money-lending classes.'[187] The less tainted *Brahmin Sabha*, with which Wedderburn had close relations, might have provided a way around this problem, but they were only willing to support the experiment under a series of conditions, which would have both undermined the purpose of the scheme and over-committed the Government of Bombay. These included interest at 9 per cent instead of 6.25; the exclusion of *ryots* currently owing more than 50 per cent of the value of their assets; liquidation of existing debts before any deal was agreed; legal guarantees that recovery of loans would take first priority; a waiver on all stamp duties and court fees and the switching of the scheme to Purandhar *taluka* in Poona, on the grounds that 'successful business could not be carried on if it were confined to a very poor taluka where security is bad, and debt excessive'.[188]

Most radically of all, the *Sabha* requested that the *taluka* chosen should not have its revenue revised for at least 20 years. The Government of Bombay gave this predictably short shrift,[189] but it raised further problems with the scheme. Stewart, having recently left Ahmednagar to become Survey and Settlement Commissioner, candidly admitted 'the exigencies of the public administration often require that poor cultivators should be dunned and compelled to pay their revenue demands by processes set on foot by the agency of the Collector'. If he was obliged to do the same to collect debts, it would seem to the 'exceedingly ignorant' *ryots* like a revenue hike, or conversely increase the risk of revenue default to that for debt repayment.[190] Ironically pre-colonial *takavi* had always been collected alongside land revenue, encouraging grants which would enhance it, and T. C. Hope of the Government of India suspected the scheme was intended to blur the distinction again, protesting:

> The more the Bank question is discussed, the more serious appear the difficulties in the way of Government interposing in what is essentially private business; and these difficulties are enhanced by the frankness with which Sir William Wedderburn and other unpractical, though benevolent, zealots now show their hand in respect of the revenue assessment.[191]

On the other hand, Buck, Hope's boss in the Revenue and Agriculture Department, was concerned that the bank could only work if changes were indeed made to the revenue system. It would need certainty that their clients never had to pay revenue in excess of their profits, as well as against 'arbitrary enhancement'.[192] As is seen in Chapter 4, the question of revenue hikes caused considerable controversy between the governments of India and Bombay. It is also noticeable that at least some officers took different perspectives on the state's right to tax as it saw fit – as well as on lenders' concerns – once the two roles became blurred. Buck clarified, 'In using the words "arbitrary enhancement" above I mean enhancement made on considerations which cannot be brought to rule. The Settlement Officer, however excellent may be his enquiry, and however just his conclusions, bases his assessment on data which the agricultural bank or the agriculturist cannot foresee or calculate.'[193]

So top-down was the whole Ahmednagar bank experiment in conception, that district officers were scarcely consulted. Rather, debates in Simla revolved around the French and Egyptian *Credits Fonciers*,[194] and 'similar institutions in Europe, America and Australia'.[195] Kimberley objected that institutions in France and settler colonies were designed for profitable farmers, and the Egyptian model had proved of little benefit to the *fellahs* (peasants).[196] A further rejoinder was sent by the Association for the Improvement of Agriculture and Amelioration for Agriculturists in Benares, who claimed to 'pity our sister Presidency thus being made an arena of experiments'. Perhaps, they suggested, the conciliators under the Deccan Agriculturists' Relief Act had been appointed 'to acclimatize that exotic idea brought from France or Siberia'. What were needed were simple, customary laws and institutions.[197] It did not help Deccan peasants that their own liberal proto-nationalists – the *Poona Sarvajanik Sabha* – were more keen to claim lending opportunities for themselves than to advocate the *ryot* cause in similar fashion. Catanach goes as far as to suggest that Mahadev Govind Ranade, one of the *Sabha*'s leaders, who also sat on the Deccan Riots Commission, had little sympathy for the 'perpetually struggling small man'.[198] Within the decade, Jotirao Phule was organising campaigns against similar agents.[199] However, no amount of state debate or restructuring of institutions could alter the fundamental problem of credit in Ahmednagar. Poor and vulnerable peasants will always be bad risks, so any lender must either have strong powers to recover his investment, or else be prepared to sustain losses. This applied as much to banks, co-operatives, philanthropists and the state *takavi* system, as to the maligned *Marwaris*. The central government's enthusiasm for Wedderburn's scheme in Ahmednagar came close to challenging the Government of Bombay's long-held assumption that indebted *ryots* were beyond help. But when it came to the bottom line, the state decided it could do nothing after all and the Ahmednagar agricultural bank scheme was abandoned without getting off the ground.

Conclusion

Rural indebtedness was the cause of considerable misery, tension and colonial debate in and well beyond Ahmednagar in this period. It was a difficult issue to unpack for the Bombay Government, because it impacted on three of their conflicting agendas: the imposition of fair and consistent legal processes, the desire to modernise the agrarian economy by encouraging investment in land (preferably by the holder himself) and the unwritten reliance on alliances to govern, including with *Marwari sowcars*. The need for some peasant borrowing was recognised, but at the same time widespread indebtedness was seen as the biggest obstacle to British hopes for the improvement of the district and its population. The tendency to see manipulative or usurious lending practices as the heart of the problem – thus creating a convenient external source of blame – prevented the state from taking a more integrated view of peasant production, poverty and vulnerability. The Deccan Riots Commission's tentative attempts to consider the broader context were lost in colonial ambivalence and internal disputes.

In some ways, the farce surrounding the proposal for a bank in Nagar *taluka* was a microcosm of colonial debates on credit in the district. Interest in the subject was only significant after the visible riots and famine crisis, and was centred on the assumption that agents other than the state were the perpetrators of all local difficulties. Not only did further discussion partially confound this view, it led to the conclusion that collaborating with the same beneficiaries of peasant poverty might be the best way to alleviate it. The aim to help Ahmednagar peasants recover from a famine that a shortage of credit – exacerbated by the state's interference – had done much to worsen was soon lost sight of in the consequent attempt to satisfy commercial concerns. In the end, although it started too late and went on too long, debate on the matter foundered because of the weakness of the original concept of an agricultural bank overseen by the state to challenge economic orthodoxies about government non-participation in any kind of market.

4
Land Revenue Rigidity, Revisions and Non-remission

Introduction

In earlier chapters, it has been argued that poor Ahmednagar peasants' chronic struggle was not fully appreciated by the upper echelons of the colonial state. Their struggle was often exacerbated, particularly through the *ryotwari* land revenue system, with which this chapter is concerned. It has been suggested that this put pressure on cultivators to take both exploitative loans and unwanted risks. Thus the fiscal interests of the state came into open conflict with smallholders' attempts to maintain their food security. It cannot be said that the weight of the land revenue was solely responsible for famine. However, in examining whether the relationship between peasants and the state enhanced or mitigated their vulnerability, it is important to focus on its most direct aspect. The land revenue system was at the heart of British administration in the countryside. District officers were, after all, called collectors. With the exception, perhaps, of the civil courts, the revenue administration was the peasants' only point of contact with the colonial state. Baden-Powell suggested that the *ryotwari* system required 'The administration ... to take a sort of paternal or "lord of the manor" interest in the whole range of agricultural conditions.'[1] While this philosophy was not much in evidence in the 1870s, the impact of the revenue system was considerable on the agrarian political economy from its conception. It is therefore necessary to consider how the land revenue affected people's lives throughout this period, as well as the reasons why its curtailment did not feature significantly in the government's response to the crisis.

Revenue rates in Ahmednagar in the 1860s and early 1870s were too light to be the primary cause of peasant poverty. However, a fixed demand in an uncertain climate added to their insecurity. There were not enough good seasons in which to save, as the system demanded, and poor ones involved higher consumption costs as well as reduced income, pushing *ryots* into dependence on borrowing. Declining grain prices also meant that the revenue demand increased in real terms over time. More significantly, rates were subject to revision every 30 years and this process was initiated in Ahmednagar shortly before the famine. Large revenue increases were anticipated and implemented during and straight after the crisis. As it was shown at the time that the largest tax rises came on the poorest land, this hike severely affected peasant viability and caused furious rows between the different branches of the state. The Survey and Revenue Departments of the Government of Bombay were constantly at odds throughout the period and, further, Ahmednagar's revenue revisions became the focus of hostile exchanges between the presidency and the Government of India, adjudicated by the Secretary of State in London. These highlighted both the uncertain principles on which *ryotwari* revenue assessments were based and the ways in which the famine crisis altered perceptions of peasant struggle. Cultivators' prospects were also harmed during the famine by a sudden change of policy on the collection of revenue when widespread crop failures made it unpayable. Richard Temple's first action as Governor of Bombay was to order suspension rather than remission of the demand, followed by aggressive attempts to recover lost revenue in the following years. In combination with revenue rises, this meant that the state's share of peasant income increased hugely in the aftermath of the famine, prolonging the period of suffering to the extent that relief had to be re-opened in Ahmednagar in 1881.

This chapter will examine the importance of land revenue as a source of state income, the history of the *ryotwari* system in Bombay Presidency and the specific ways that its rigidity exaggerated the burden on the peasantry. It will also explore the basis on which revisions of assessment were calculated, contemporary proposals for a more moderate system and the particular history of revenue increases in Ahmednagar between 1875 and 1884, before investigating the effects of the non-remission policy during the famine and the subsequent collection of arrears.

The importance of land revenue

Famines occur in rural areas. Capitalist exploitation, social marginalisation and state neglect can all be found to large degrees in cities, yet it is food-producing areas which have always been more vulnerable to failures of supply. It is therefore significant that land revenue was the largest source of taxation in India in this period. The Famine Commission Report calculated that the gross revenue of the Government of India in 1878–79 had been just over £65 million, of which £24 million 'may be regarded as in no proper sense raised by taxation'.[2] This included fines, opium duty and income from government services and productive works. Of the remainder, £22.5 million was land revenue. It was argued that even this was not 'taxation proper' on the grounds that it represented a share of agricultural surplus to which the state was entitled as rent.[3] This compared with land revenue income of £14.58 million in 1840, the rise being attributed primarily to acquisition of new territory and extension of cultivation.[4] The Bombay Government, too, acknowledged in 1875–76 that 'the chief source of revenue in this Presidency is the land'.[5] In Ahmednagar, a higher proportion of district income – 41 per cent – came from the land revenue, despite the lack of investment in peasant production.

The presidency total had increased that year by £6,280 to £3,694,356 owing to revision settlements, but was lower than it had been in 1872–73, at £3,751,050.[6] Tellingly, this was explained by 'the relinquishment of land chiefly in Ahmednagar, Surat and Kaira'.[7] Throughout the 1870s this had largely cancelled out – and perhaps from the state's point of view increased the need for – significantly increased revenue settlements. Under these circumstances, some concern was expressed at the heavy reliance on land revenue, which was becoming an increasingly regressive form of taxation. The Secretary of State, Salisbury, graphically suggested in 1875 that 'as India must be bled, the lancet should be directed to those parts where the blood is congested, or at least sufficient, not to those which are already feeble from the want of it'.[8] Pedder then calculated that declining produce prices at the same time as settlement revisions had greatly increased the proportion of agricultural income liable for tax. Table 4.1 gives his averages for the Deccan districts.

Revisions continued apace – only starting in Ahmednagar in 1875 – as prices fell and famine struck. This reflected problems for the state, too, in depending on land revenue during a depression. It was not only the largest source of income but also the easiest to raise. Land revenue rates only began to decline in real terms and in relation to other taxes and receipts in the early twentieth century. Land revenue fell from a national average still up at 40 per cent of state income in 1900 to around 25 per cent in 1909 – co-incidentally the period in which, according to Bhatia, the famine problem throughout India diminished.[9] At this point, it was declared that 'an increase in revenue in future shall be looked for

Table 4.1 Incidence of land revenue as a percentage of produce value, Bombay Deccan, 1830–75

Period	Stage of settlement	Average price of produce (*seers* per rupee)	Estimated value of produce per acre (rupees, *annas* and *pice*)	Average rate of assessment per acre (rupees, *annas* and *pice*)	Assessment as percentage of value of produce
1830–37	Before first settlement	43	4-13-5	1-0-6	22.17
1837–48	First 10 years of settlement	56.5	3-8-7	0-8-4	14.7
1862–67	Last 5 years of settlement	18	11-1-9	0-7-5	4.17
1870–75	First 5 years of revision settlement	27.5	7-5-5	0-11-9	10.0

Source: FCR (1880), Appendix I, p. 158.

rather in improving trade and industry than in adding largely to the burden of the land'.[10] Attempts to find alternative forms of taxation earlier, however, were unimpressive.

Income tax was introduced in 1868 but caused confusion because land revenue payers were not automatically exempted. It proved so impossible to set a uniform tax rate on varying agricultural profits that the Government of Bombay was permitted in 1870 to exempt all those earning less than Rs 125 per annum, and Rs 500 per annum in districts where the land revenue settlement was undergoing revision. Nonetheless, a threefold increase saw Rs 79,738 of income tax levied in Ahmednagar district in 1870–71.[11] Following considerable protest, the threshold was raised the following year to Rs 750, and the rate reduced from 3.12 per cent to 1.04 per cent (two *pice* in the rupee). It was claimed that this left the tax 'entirely free from the objections that have been so constantly urged against it', but a sample survey of Belgaum district revealed a mere 973 payers, mostly moneylenders, some traders and 171 landholders. Even in Bombay city, the tax fell on just 4 per cent of *Parsis* and 1.5 per cent of Hindus (as well as 39 per cent of Europeans).[12] In 1872, the threshold was raised once more, to Rs 1000, reducing the total revenue in the Presidency to 1.1 m rupees, after which the Government of India abolished income tax entirely as too unpopular.[13] Not only did this again increase the relative tax burden on rural areas, it made it impossible to tax agricultural traders and moneylenders, 'who, and not the ryots, are the chief gainers in years of prosperity'.[14] The Government of Bombay protested strongly that a moderate, well-targeted tax that suited a large rich city and had overcome 'slight opposition' elsewhere should be denied them.[15] Their concern was shared by Temple who, as outgoing Finance Member of the Viceroy's Council, fought unsuccessfully to retain income tax. This was one of his first policy battles with his successor, John Strachey, who saw it as 'neither politic nor just'.[16] Income tax was permanently re-introduced soon after Strachey's own retirement, in 1881, when Lord Ripon replaced Lord Lytton as Viceroy. Thus, despite efforts to target it, and a local consensus that it reasonably raised revenue from those best able to pay in the presidency, the Government of Bombay was unable to levy income tax from 1873 to 1881.

Other forms of taxation presented a similar story. Those that fell on wealthier groups than cultivators were contested more effectively as a result of their greater access to the state. They also often took the form of colonial experiments in taxation, lacking the historical legitimacy of land revenue, and sometimes contradicting the logic of non-interference in commercial activity. The minimal trade tax, for example, was suspended in Ahmednagar during the famine, to relieve 'the pressure on the trading classes' caused by the scarcity.[17] This seems extraordinary given the non-remission of land revenue and contemporary perceptions that local traders were profiting immorally from the famine. It was consistent, though, with the reliance on free trade to meet the demand for food and Temple's unsubstantiated assumption that peasants were able to cope with the crisis without assistance. Another attempt to find new targets for taxation was the non-agricultural cess, levied in 1871–72 on 'every person being a head of a household ... who ... carries on any profession, trade, or calling other than the cultivation of land'. The rate was $1\frac{1}{2}$ rupees for those earning under Rs 100, three rupees for an income up to Rs 500, and six rupees thereafter, with those paying police charges in Bombay city and suburbs exempted.[18] A tax that fell largely on the urban poor was immediately unpopular, especially starting in another year of widespread scarcity, and the cess was abandoned within a year, along with the police rates. This proved a watershed for the Bombay Government, with the consequent loss of Rs 666,192 resulting in an excess of expenditure over income of Rs 354,520. Rightly anticipating criticism from above, they protested at their lack of access to resources:

> As every effort has already been made to conduct the administration with the greatest regard to economy, this Government have reluctantly

come to the conclusion that the means at their disposal are insufficient to meet the legitimate demands of the several departments included under Provincial Services. They have, therefore, found it necessary to bring the present position of Provincial revenues prominently to the notice of the Government of India.[19]

Thus other taxes than land revenue failed to meet the state's needs. Moreover, with the exception of the short-lived non-agricultural cess and municipalities' income, which came from *octroi* and a deeply unpopular house tax, the peasantry was additionally liable to them. For example, scarce Local Funds for district works were funded primarily from the one-*anna* cess, an extra levy on land revenue paying land. With income tax disallowed and land revenue no longer growing through expansion of cultivation, there was enormous pressure both to raise land revenue rates and reduce spending in Bombay in the 1870s.

This was despite the fact that Bombay rates were already high relative to other presidencies. As land revenue was an imperial tax, much of it was channelled into the central treasury. Bombay's contribution came at a rate of six shillings *per capita*, including non-taxpayers. This compared with four shillings – the national average – in the other *ryotwari* presidency, Madras, and just three shillings $2\frac{1}{2}$d *per capita* in the more populous permanently settled provinces of Bengal and Assam.[20] Further, the proportion of land revenue to the estimated value of aggregate produce came to 7.6 per cent in Bombay, higher than anywhere but the North West Provinces,[21] where the average charge amounted to 30 per cent of the value of the land, compared with 80 per cent in Bombay.[22] Bombay's belief that it raised more than it was allowed to spend caused tension with the Government of India, for whom Strachey responded by criticising Bombay's financial management. The contested priority of national interests over local was highlighted during the 1874 famine in Bihar. A fall in Bombay's balances from Rs 43,031,200 to Rs 25,808,610 in 1873–74 was attributed to 'large remittances to Calcutta and Patna, in order to strengthen the treasuries in the famine districts'.[23] Ironically, the conception of famine as a national responsibility had been diminished by the time of Bombay's own crisis, as is seen in Chapter 5.

Despite the Bombay Government's complaints at its revenue contributions, its Survey Department proceeded in the 1870s to increase them still more disproportionately. In the round of revisions up to 1883, Bombay land revenue rises averaged 32 per cent, compared to 14 per cent in the North West Provinces and 7 per cent in Punjab. The latter figures reflected a reduction in the proportion of each cultivator's wealth it was felt safe to demand, counteracting the levy on improved prosperity.[24] Bombay's failure to do this was repudiated in 1879 by Secretary of State Cranbrook who argued that increases to the cultivators' burden in the Deccan in 1877–78 had been higher 'than has been found practicable in other Provinces'.[25] Even the Government of India recognised in the aftermath of the famine that 'the new assessments in Nevasa taluk of the Ahmednagar Collectorate introduce a much higher increase than in the north of India would be considered safe'.[26] Wider controversies surrounding Ahmednagar revenue revisions are considered below. As far as the Bombay Survey Department was concerned, it was more meaningful to compare their rates with the previous settlement by Sir George Wingate, which had been too light, than with those in other provinces. It is therefore useful to examine the origins and basis of the land revenue system in the presidency.

The history of the Bombay *ryotwari* assessment

The British aim in imposing the *ryotwari* revenue system in Bombay Presidency had been to secure a fixed return, removing the perceived arbitrary character of the *Maratha* system which levied a proportion of the annual harvest, as well as the abuses of the *zemindari* system they had created in Bengal. It was believed that, by taxing each cultivator directly, rates could be kept to a minimum. Agricultural development would also be encouraged by guaranteed security of tenure provided the annual demand was met. Such a small amount of the profit of a good year would be taken that successful peasants would be able not only to save as a precaution against less fruitful seasons but also to invest in their holdings. A similar benefit would be felt from the retention of set rates for 30-year periods. The *ryotwari* settlement involved the measurement and classification of every single field, with a scale being constructed to ensure fairness for all qualities of holding. The basis for assessment was David Ricardo's theoretical Law of Rent, which sought to link the revenue to the – initially notional – value of the land, ensuring lower charges in less successfully capitalised areas. The idea of charging a rent rather than a tax rested in turn on John Stuart Mill's idea of the state as a 'universal landlord'.[27]

Eric Stokes argues that this thrust the government into every sphere of Indian life, but in a disguised way. The combination of 'light taxes and good laws' eschewed direct management of the economy, but Mill's doctrine hid a great reform agenda: 'It meant using law in a revolutionary way, consciously employing it as a weapon to transform Indian society by breaking up the customary, communal tenures.'[28] Taxing cultivators individually rather than collectively followed naturally from the commercialisation of agriculture, accentuating the vulnerability of poorer farmers. Similarly, Ravinder Kumar contends that after the first Governor of Bombay, Mountstuart Elphinstone's, attempt to preserve traditional social and revenue institutions, his utilitarian successors implemented an atomising tax philosophy which led to the deliberate polarisation of the peasantry in both social and economic terms.[29] Famine Commissioner Sullivan also condemned the theory of rent as a licence for exploitation, including the imposition of extra cesses on peasants – who saw no difference between any state levies – on the spurious grounds that they did not already pay tax. As legislation had frequently been passed to acquire land for public use, he declared, the claim to be universal landlord was 'a doctrine for which I believe there is no historical foundation, which the action of the Government itself goes to disprove, and which if accepted might lead to most mischievous results'.[30]

The first Bombay settlement was completed in 1836 by R. K. Pringle, a recent star student from Haileybury, whose reliance on theory resulted in experimental rates equivalent to 55 per cent of the value of an average crop. This proved disastrous, the Government of Bombay subsequently acknowledging that 'experience had shown that those levies were too exhausting to a cultivator in the condition to which agriculture had sunk'.[31] The subsequent Survey Commissioner, Wingate, went further, concluding, 'There can be little doubt that the over-estimate of the capabilities of the Deccan formed and acted upon by our early Collectors, drained the country of its agricultural capital and accounts in great measure for the poverty and distress in which the cultivating population has ever since been plunged.'[32] As a result, Pringle's assessment was quickly abandoned in favour of Wingate and Goldsmid's more moderate rates, determined on the basis of a full survey, taking over 15 years. This meant that the introduction of their settlement was staggered. Thirty-year rates were set in Ahmednagar between 1845 and 1855. Wingate still based his assessment on the theory of rent, hence the attention paid to the condition of each field. However, his liberal recognition of *ryots'* difficulties and disinclination to charge for potential rather than actual farming income allowed for considerable agricultural growth, for which the government was pleased to take credit, although it mostly resulted from expansion rather than intensification of cultivation.[33]

As seen in Chapter 2, Sumit Guha argues that this expansion ultimately proved harmful to the agrarian economy of the Deccan, and thus that low taxation at this time was not wholly beneficial.[34] That does not mean, however, that higher revenue rates imposed in the 1870s were reasonable. McAlpin, on the other hand, argues that the Bombay Settlement ultimately achieved its aims of providing the stability and incentive to allow agricultural growth as well as cutting out the rapacious middleman of the *zemindari* system.[35] While this may be true of the period of Wingate's settlement, it is questionable whether it remained so after its revision. McAlpin seeks to refute the link, suggested by Dutt and Bhatia, between land revenue and nineteenth-century famine in Bombay Presidency by demonstrating that its incidence fell in real terms.[36] It is true that over the whole period of Wingate's settlement and its first revision, it rose by less than the price of *jowar*. Famine vulnerability can not, however, be quantified from long-term averages. As seen in Chapter 2, prices fell steadily between 1870 and 1884, except during the period when they rose in response to the scarcity itself. Moreover, while Wingate's rates were indubitably light, the view that they were generous towards peasants was the very justification for the large increases seen during the period of study.

In the light of the role of moneylenders seen in Chapter 3, McAlpin's further assertion that the *ryotwari* system also broke the hold of agrarian elites who had held back their fellow cultivators is even more contentious.[37] Neil Charlesworth argues that progress was achieved in Deccan agriculture in the nineteenth century precisely through stratification of poorer and richer peasants. Moreover, it reflected elites' transport advantages in gaining genuine access to the market, not the revenue system.[38] In Ahmednagar, lack of growth in peasant cultivation makes both views optimistic. As Neil Rabitoy and David Washbrook have argued, local elites were indeed undercut by individual assessment, but the benefit was for the state.[39] Rabitoy asserts that *ryotwari*'s theoretical justifications only evolved later out of the expedient objective of maximising revenue while maintaining political stability. This was achieved through negotiation, in which some local elites were rewarded with government positions or *inam* (low rent) lands, but others were deliberately weakened as state and system became more institutionalised. Thus the structure of direct taxation was 'incidental, almost accidental' to the aim of establishing productive control over the rural economy.[40] Charlesworth suggests that Deccan *patels* were undermined in the earliest stages of British rule, leaving them politically powerless to resist the 'purest ryotwari' at harsher junctures, such as Pringle's oppressive settlement. Only more robust elites in richer districts in the Konkan or Gujarat successfully demanded mitigation.[41] Whether or not it was successful in transforming peasants' prospects, the *ryotwari* settlement was thus characterised by its lack of constraint on the state's tax-raising powers on land.

At the turn of the century, Baden-Powell hinted that the Bombay Settlement remained a misnomer, as 'settlement' implies discussion at the village level, whereas Bombay's 'almost scientific principles' gave occupants no choice other than to pay or leave.[42] The Bombay Government pointed out in 1873 that 'The more intelligent among the agricultural class were even taken into consultation on the subject, and their opinion allowed due weight' during Wingate's survey,[43] but their tone of surprise did not bode well for Ahmednagar peasants when Wingate was not around. The lack of consultation was later highlighted by the nationalist Romesh Dutt, who noted the irony that direct taxation had proved more burdensome than the Permanent Settlement in Bengal, where *zemindars* took an extra cut of the revenue. At least there, he argued, the state sought to limit taxation and guaranteed it would not rise, whereas self-regulation inevitably inclined the Bombay Government to increase it without due caution.[44]

Cultivators had also protested with occasional success against the pressure of the land revenue in the early years of British rule in Bombay,[45] but this had become considerably harder by the 1870s. When Elphinstone created the Bombay revenue system, he instituted the right of petition, and of civil courts to adjudicate in cases between government officers and the population, with the offence of 'Undue Exaction of Revenue' punishable by up to seven years in prison.[46] Though such extreme sanctions were not used, civil courts upheld appeals against the revenue demand as late as the 1860s, leading to accusations that some judges fancied themselves as 'a bulwark against the tax-gatherer'.[47] However, following 600 appeals in the early 1870s, which the Government of India perceived as 'rather ... political than legal action',[48] jurisdiction over revenue matters was removed from the courts, increasing *ryots'* frustration in advance of the Deccan Riots. Governor Wodehouse himself admitted to the reservation that 'prohibition of enquiry seems to me improper'.[49] Eight years later, when Ahmednagar Assistant Collector Anding reiterated Hamilton's suggestion that *ryots* should have a right of appeal against *mamlatdars'* decisions on the annual demand, he was accused of showing 'more boldness than discretion in repeating' such an outlandish idea.[50] Meanwhile, Bombay newspaper *Jame Jamsed* complained that the existing petition system deliberately neutralised complaints by burying them in bureaucracy.[51]

Difficulties in meeting the land revenue demand

The problem with the *ryotwari* revenue system in Bombay Presidency was not that it was too high, at least under Wingate's settlement, but that it was applied rigidly in uncertain conditions. McAlpin typically echoes the state's own case that 'clear and constant expectations about revenue collections' reduced the insecurity of cultivation,[52] but the same inexorable constancy was more of a burden than an incentive for poorer peasants. The Deccan Riots Commission Report noted that 'a Revenue System which levies from the cultivators of a district, such as that now dealt with, the same amount yearly without regard to out-turn of the season, must of necessity lead to borrowing. In bad years the ryot *must* borrow'.[53] This recognition that moneylenders were often required to pay the revenue – and indeed that their unwanted emergence had partly been facilitated by the revenue system – did not encourage any thoughts of reduction, however.[54] Rather, the Government of Bombay held that if the exploiter was paying anyway, land revenue rates made no difference to the impoverished *ryots*, and indeed that higher charges would be an effective way of penalising the middlemen.[55] As Cranbrook pointed out, this justification was inconsistent with the aims of the Deccan Agriculturists' Relief Act.[56]

Baden-Powell attempted to essentialise *ryots'* difficulty while acknowledging that 'there is a rigidity to our system that, whatever its justification, is not always acceptable to the Oriental mind'.[57] In a footnote, he recognised that the problem was broader, quoting Holt Mackenzie from the 1830s: 'men, especially men so improvident as the natives of India, do not live by averages'.[58] The only flexibility within the annual demand was the *annewari* system, under which a good crop was graded at 16 *annas* and an average one at 12. Though crop assessments could be negotiated by holders, the *kulkarni* or *mamlatdar* made the decision on behalf of the state and was under pressure to maximise revenue results. Moreover, they served only as a guide to remissions or suspensions, with no guaranteed reductions. In the early twentieth century, F. G. H. Anderson calculated that the whole arbitrary scale was flawed. On the poorest land, the average yield relative to a good season was six *annas*, with revenue becoming unpayable without borrowing or other employment at four *annas*.[59] The Deccan Riots Commission Report concluded that a system of fixed demands and occasional remission was as inefficient for the state as it was unreasonable for cultivators of poor land:

> We do not at all under-estimate the importance of fixity of demand in the land revenue, but we question whether this advantage is not

purchased too dearly by the ryots of a large portion of the disturbed district, perhaps also by the Government itself: for the Government limits its assessment in consideration of bad seasons, but is nonetheless forced to give remissions in years of drought.[60]

The Government of Bombay responded only by removing automatic remissions in 1877, as discussed later, insisting 'everything affecting the security or insecurity of agriculture in the tract under settlement' had been taken into account in the first place, including the variability of the monsoon.[61] Survey Commissioner Colonel Francis was reported to have responded to a petition by *ryots* in Karmala *taluka* of Sholapur district 'that to make the rain fall seasonaly is in the hands of God, and the rates levied must be paid'.[62] The Government of India privately recognised that the problem was chronic but only after the famine and only in Ahmednagar: 'The local peculiarities of the district cannot be ignored; but at present they are recognised only in extreme cases.'[63]

That debates over the impact of the *ryotwari* system on peasants should come to a head during Francis' first revision settlement emphasised another feature that reduced their security – the 30-year revision process itself. Fixed revenue rates were designed to enable saving as profits increased, but the opposite effect was felt when a revision of settlement was pending. Anticipation of an unspecified increase in outlay accentuated *ryots*' aversion to risk and reduced the availability of credit. Further, as evidence from Ahmednagar will confirm later in this chapter, the assumption that rates had been too low in the last decade of the previous settlement created a temptation to recoup the state's losses when revising them. It could also seem reasonable to set an initially unaffordable rate on the basis that it would even out over the full term. Yet any large increase, implemented without gradations or much advance warning of its scale, amounted to an economic shock. For all the importance attached to annually fixed demands within each settlement, the revision process introduced unpredictability into the heart of the *ryotwari* system. As Dutt declared, '*uncertainty* is a greater evil than over-assessment'.[64]

The annual revenue demand was rigidly levied not only in its amount, but also in its timing, which did not allow for flexible strategies, such as delayed sowing in the event of a late monsoon, and prevented cultivators from withholding their stocks until the market was more advantageous. Although the precise dates and proportions of the twice yearly instalments were set to reflect each *taluka*'s preference for *kharif* or *rabi* crops, no allowance was made in Ahmednagar for more local – or individual – variation, or for changes in a given year. The Government of Bombay had recognised in 1865 that 'it is doubtful whether it would be possible, without occasioning hardship and inconvenience, to fix the instalments with reference to the prevailing crops of entire Talooks, instead of, as at present, with regard to those of particular villages'. However, they allowed it so long as 'Collectors are satisfied that it will not cause loss or injury to the ryots,'[65] which D'Oyly was in Ahmednagar.[66] This rigidity came to present chronic difficulties for some *ryots*. For example, Nagar was classed as a *kharif taluka*, growing mostly *bajri* in the early season, and paying revenue on the 10th of January and March. In 1884, Elphinston forwarded complaints from two villages there, Hingangam and Hamidpur, that 'it is now 5 years since any kharif crops were produced in these villages. The ryots complain that they are much inconvenienced owing to the dates of the instalments being fixed at a season when they cannot afford to pay their dues.'[67] Such cases were numerous. Even within villages, different crops were often grown on larger and smaller holdings, with some growing quicker than others sown at the same time. Now, however, Revenue Commissioner Robertson emphasised 'the desirability of taking general and regular action instead of treating the subject piece meal'. He suggested that leeway could be created instead by collecting revenue in four instead of two instalments.[68] This was predictably unpopular with local officers.

Instalment dates created such complications because they were deliberately set very soon after the anticipated harvest. This was to prevent *ryots* from spending their earnings before the *kulkarni* arrived, and thus to remove the need for borrowing to meet the demand. Whenever the harvest came after the due date, however, the effect was precisely the opposite. The Famine Commission suggested that the main instalment should relate to the crop that was intended for sale, as the first harvest was often for cultivators' own use. It should also be set to allow plenty of time for reaping and sale, so that *ryots* were not forced to the market at its lowest point, or even before their crops were ripe.[69] Behind the policy of encouraging peasants to pay the revenue as quickly as possible lay the Bombay Government's own constant desire for resources. Rather than allowing extra time for poorer *ryots* to meet the demand, it accepted a Government of India proposal to experiment with discounts for paying revenue in advance.[70] This had the added advantage of providing a further excuse not to suspend revenue when a poor season followed a good one, as *ryots* had had the chance to pay for up to three years from their previous profits.

The fixed tenure system also tied peasants to particular plots of land, as seen in Chapter 1. The ban on temporary migration from holdings in bad years was entrenched by a resolution in 1874 that defaulted land should no longer be auctioned but kept for grazing, in response to fears that peasants were colluding to avoid revenue.[71] The Government of Bombay alleged that friends of defaulters refused to bid against each other, bought holdings at a pittance, then leased them back in better seasons. It promised to show defaulters that they 'cannot hope to get their holding restored', adding that Collectors should be sure to strip them of all moveable property before proceeding to a sale.[72] The principle of guaranteed tenure if the revenue demand was met thus also constituted a threat of eviction for non-payment. State sanctions for revenue arrears began with a notice from the *kulkarni*, for which a fee was charged, followed by the distraint and sale of moveable property. A late first instalment also rendered the entire annual demand immediately due, with a quarter added as a fine to cover the costs of the coercive process. If all this failed to procure a payment, the holding could be sold. It was claimed that these measures were merely to ensure punctuality, and to enable recovery of assets in the event of default. Yet each financial penalty or charge enhanced the state coffers at the expense of those least able to afford it.

Sanctions for revenue arrears also amounted to a means of control over the peasantry, discouraging protest against rates and legitimising the replacement of unprofitable farmers. Havelock recorded that new rates had only been paid in Madhe *taluka* of Poona district after 'preparation for rigid enforcement of distraint and eviction' had put down a peasant boycott.[73] Ahmednagar Judge Wedderburn claimed more seriously that a perceived 'combination among the ryots to resist the Government demand in expectation of obtaining a reduction in the revised rates' in Bhimthadi *taluka* explained a sharp rise in actual sales for arrears between 1874 and 1877.[74] The Bombay Government maintained that the threat of eviction had been sufficient to force the offending peasants to back down, explaining the rise in sales as the result of 'failure of crops and the impoverished condition of the ryots'.[75] Perhaps then, in the face of revenue hikes, non-payment due to protest and poverty could only be distinguished from each other by the response to threats, allowing no room for sympathy at poor harvests. This only added to *ryots*' desperation. The *Poona Sarvajanik Sabha* pointed out that the 17 Bhimthadi villages where distraint had been greatest were at the epicentre of the Deccan riots.[76]

Evictions were said to be a rare last resort, but the cost of the threat itself further reduced the viability of household economies. Table 4.2 shows the various sanctions taken for arrears in Ahmednagar. A picture can be seen of extensive and cumulative pressure to pay the revenue – in

Table 4.2 Sanctions for revenue arrears, Ahmednagar, 1873–82

Year	Total occupants	Notices served		Interest		Distraint		Realisations (rupees)		Occupancies transferred	
		Cases	% to population	Cases	Amount due (rupees)	Cases	% to population	Sales of property	Sales of land	Cases	% of occupancies
1873–74	na	7,173	na	859	40,596	44	na	110	1,283	na	na
1874–75	na	2,023	na	256	5,814	130	na	201	279	na	na
1875–76	na	3,028	na	383	11,898*	161	na	242	328	na	na
1876–77	na	17,621	na	8	17*	65	na	0	1	na	na
1877–78	na	36,110	na	34	330*	59	na	351	54	na	na
1878–79	na	27,870	na	48	925*	226	na	725	3,606	na	na
1879–80	64,571	15,591	23.18	129	1,014*	621	0.96	2,195	13,845	1,237	1.91
1880–81	64,716	16,922	26.14	84	225*	141	0.21	223	812	1,282	1.98
1881–82	64,500	15,086	23.40	5	1	165	0.25	68	910	426	0.66

na: not available.
* From 1875–76 to 1880–81, the fine of a quarter of the demand was recorded instead of calculated interest.

Source: Boswell to Havelock, No. 1645, 20 July 1874, p. 127; Boswell to Oliphant, No. 2132, 20 July 1875, p. 476; Boswell to Havelock, No. 1952, 20 July 1876, p. 239; Jacomb to Robertson, No. A/4960, 19 July 1877, p. 33; Stewart to Robertson, No. 3195, 22–4 July 1878, p. 306; King to Robertson, No. 3140, 19–23 July 1879, p. 37; King to Robertson, No. 4161, 20 July 1880, p. 47; King to Robertson, No. 4584, 22–5 July 1881, p. 99; Ephinston to Robertson, No. 5730, 20 July 1882, p. 139.

the form of interest charges and sales – in the years of depression that built up to the famine crisis. Sales for unofficial departures as a result of the famine were recognised only in 1879–80, after which enforced sales continued in greater numbers than before. The 1879 Bombay Land Revenue Code was said to contain 'much more stringent provisions against tenants than the code of 1827 formerly in force'.[77] Their use was at the discretion of collectors, however, and the degree of pressure varied over time and between districts. For example, Robertson noted in 1880–81 that 'the issue of notices was found to be necessary to a greater extent in Ahmednagar, Poona and Nasik than in the other Collectorates of the [Central] Division'.[78] The impression is that pressure to recover the land revenue in Ahmednagar was deliberately harsh after the famine. Noting that the value of land sales in the Central Division in 1883 was 163,187 rupees, whereas sales of other property for arrears brought in only 4,426, Secretary of State Kimberley enquired whether the Bombay Government was seeking to get rid of pauper occupants, to be replaced by wealthier cultivators. He did not oppose such a policy in principle, but warned that it was liable to lead to revenue losses where demand for land remained low, adding that the poverty of many cultivators must have been partly the result of excessive revenue pressure during and directly after the famine.[79]

Calculation of the revised revenue settlement

The precise criteria for assessing the *ryotwari* revenue demand were never spelled out. Buck, of the Government of India, complained in 1882 that it was hard to judge 'the severity of the assessment in certain parts of the Bombay Presidency' because 'There are no principles for calculation of assessment rates in the [Bombay] Revenue Handbook.'[80] Moreover, 'no principles for the assessment of land revenue have been enunciated by the Government of India'.[81] Levies were set after enquiries into the size and soil quality of individual fields and took into account increases since the last settlement in, variously, produce prices and yields, land values, the extent of cultivation, population, agricultural stock, irrigation and transport facilities. It was often claimed that such extensive and detailed enquiries meant that the survey process was a scientific one, and its outcomes therefore justified.[82] However, only soil quality was graded according to a fixed scale, with the other factors used to determine the rate of increase to be applied to each band within it. Thus the actual rate of levy remained arbitrary. Even Baden-Powell, who regarded the experience of survey officers as sufficient to guarantee 'the greatest accuracy and fairness', acknowledged that

> Settlement is very much a matter of individual taste and opinion, and ... the elaborate tables and calculations do not produce much but expense and long report-writing. There is, no doubt, in every assessment, a point where it comes to taking a certain figure, which implies an element of personal judgement – the intuitive conclusion of a trained mind accustomed to the work.[83]

Regardless of the quality of the personnel, in an era when the government was keen to adopt scientific methods rather than drawing on local revenue officers' own experience and opinions, it was all too easy to exaggerate consistency and conflate it with fairness. Survey Commissioner Stewart admitted otherwise: 'A good classification in itself will not prevent the imposition of too high or too low an assessment, but it is a complete safeguard against relative inequality of assessment.'[84]

In theory what survey officers strove to estimate was each holding's rent value. In spite of the condemnation of Pringle's attempt to apply it in Bombay, Stokes argues that 'the rent doctrine had triumphed, in large measure, in the forming of the assessment theory and method of the land revenue systems of India'.[85] Even Goldsmid and Wingate declared their belief 'in the justice of the principle of limiting the Government demand to a portion of the true rent' but admitted 'the difficulty of ascertaining [it on] different descriptions of land'.[86] Thus, while Temple proclaimed that 'the present system of fixing the rents in money is economically scientific, is civilised, is worthy of the British Government, [and] is distinctly conducive to all those moral qualities which bring a nation forth out of barbarism',[87] problems in carrying it out had been recognised by the creators of the Bombay Settlement. In an impoverished district like Ahmednagar, estimating what share of agricultural income was neither reasonable profit nor production costs – without enquiries into personal finances – was thankless, and liable to produce unsatisfactory results for revenue purposes. The more common estimation of the true value of land was not much easier because land markets were heavily distorted. Throughout the 1870s in Ahmednagar, resignations of holdings created such an excess supply of land that many had little market value.[88] As early as 1866, the Bombay Government approved a suggestion that revenue officers should themselves bid on behalf of the state when land was auctioned for non-payment, to prevent them being sold for less than the value of the arrears.[89] On the other hand, land sales in fulfilment of debts were recorded for exaggerated amounts, as seen in Chapter 3. These notional high land prices were often regarded as a legitimate reason to increase the revenue demand.[90] However, market values were not taken as read. The Survey Department attempted to estimate rent on the basis of its circumstantial enquiries.

A particular justification for increasing the ascribed rental value of holdings was their proximity to state-financed infrastructural works. A nearby road or railway that had not been there at the previous settlement could be assumed to generate greater potential profits by improving access to markets. Whether such profits could be realised by *ryots* or not, the government felt entitled to recoup a share of its expenditure. There was also an extensive debate about the extent to which revenue should be enhanced on account of the construction of irrigation works. After disputes between the Irrigation and Survey Departments as to which should receive the proceeds of such taxation, it was decided that holders should be charged a 'Protection Rate' on the potential benefit of proximity to canals, in proportion to the state's outlay, rather than on any measure of their actual effect on cultivation.[91] Even if landholders took no water, the 1879 Bombay Irrigation Bill proposed that they should pay for 'benefit from percolation'.[92] This was rejected by the Government of India, but they advised that canal charges could be made as part of the land revenue, as the state had the right to share in the increased profits of cultivation engendered by its own efforts.[93] This principle of enhancing the revenue to take a share of potential profit remained, despite Secretary of State Lord Hartington's warning that canals depending on rainfall for their supply had not been proved to be beneficial in the Deccan, 'where irrigation on a large scale is still a novel and doubtful experiment'.[94]

The most quantifiable method of estimating rent was produce prices. The Government of Bombay argued that 'value may be judged by examining the scale of prices over a long period and noting the proportion of increase which appears to be permanent'.[95] Indeed it was suggested that Bombay's revenue increases were higher than elsewhere because they used prices instead of the concept of natural rent as their main guide.[96] This was a problematic rationale for hikes in Ahmednagar in the late 1870s and 1880s. While prices were higher than they had been at the time of Wingate's settlement, the trend had been downwards since the 1860s, culminating in lower prices in 1874 than at any other time in the nineteenth century apart from the slump following Pringle's settlement, from 1837 to 1846. The Government of India argued that taking a 30-year average could cause hardship:

> In view of the great fall of prices and the vicissitudes of season in the Deccan during the last few years, it would be desirable that

the present Government of Bombay should consider whether the recent ... revisions of the revenue have given sufficient relief from an assessment which was based, in part, on an unduly high estimate of the normal value of field produce in the Deccan.[97]

Moreover, focusing on the profits of cultivation, however crudely, equated to a tax on agricultural income, and Baden-Powell was to observe that in that case land revenue should logically be replaced by a broader income tax.[98] Furthermore, not all crops were ever sold, and if consumption was charged for as well as sale, smaller peasants would lose the greatest proportion of their income. Hartington chastised the Bombay Survey Department for persistently presenting him with evidence that revenue rises trailed behind the rising value of crops, when 'an enhancement in arithmetical proportion to a rise in prices is not necessarily, as you assume, a just one'.[99]

Nonetheless, in the context of constantly changing produce prices, many argued for a return to a levy calculated annually according to incomes, as well as for other flexible systems of revenue assessment. Wedderburn suggested fixing the revenue at one-sixteenth of the sale value of crops. Anticipating the frequent argument that a fixed sum afforded certainty and convenience to payers as well as collectors of revenue, he proposed to give cultivators the option of paying three-sixteenths of the average yield at average prices, regardless of actual out-turn or market rate.[100] This was more practical than calls for a return to the *Maratha* system of taking a proportion of annual yields in kind,[101] though the Government of Bombay acknowledged that this was how most *ryots* dealt with moneylenders, and that more account might be taken of the scarcity of currency in circulation.[102] Assistant Collector Hamilton suggested that an annual assessment could be based on seasonal rainfall.[103] Though he got short shrift from Robertson, the Government of India asked the Government of Bombay to consider, at least,

> Whether in these four districts [of the Deccan], or in parts of them, it would not be wise to have a varying scale of revenue demand to be applied in unfavourable seasons, whereby the normal assessment might be reduced by a certain percentage over an entire district, or division of a district, in the event of a failure of rain or other cause of serious damage to crops.[104]

This implied a two-tier revenue system, whereby bad seasons and more vulnerable regions attracted different rates in an effective formalisation of the remission system, or else the delegation of greater authority over revenue issues to district officers. When commenting on revenue revisions in Newasa *taluka*, Buck suggested that Ahmednagar was exactly the sort of district where such ideas should be tried: 'I question whether the system of fixed rates is a sound one unless accompanied by strict provisions for remissions or suspensions in years of drought.'[105] As seen above, however, the Bombay Survey Department maintained that their rates already took account of poor seasons, so extra provision for remissions was unnecessary. Yet there was inconsistency in their argument. Havelock pointed out that Francis had claimed in 1874 that his high revision rates in Madhe *taluka* would work with the help of sensible remissions in hard years, having argued strongly against remission in the scarcity of 1871. His views showed 'a change, at least in the application of the theory, so considerable, that it must be regarded as a complete recantation'.[106] Baden-Powell suggested that revenue flexibility in poorer regions might best be achieved by increasing collectors' powers to remit and suspend revenue in poor seasons.[107] Though it was easier to make such a case in the 1890s than the 1870s, Wedderburn considered the notion of an entirely free assessment, to be made by local revenue officers every year,[108] and Hamilton's levy according to rainfall would have worked in the same way. Stewart condemned the idea as uncertain, depending as it would on 'the varying views of individual revenue officers or even of successive Governments' and likely to expose officers to undue influence.[109] Yet this was no less true of Survey Department officers. His views reflected a battle for supremacy between them and the Revenue Department.

A more popular and simple suggestion was that the Deccan should be permanently settled. This had been proposed by Baird Smith after the 1860 North West Provinces famine, with the initial support of Viceroy Lord Canning and Secretary of State Sir Charles Wood, and was not formally rejected until the 1884 Bill to Amend the Land Revenue Code.[110] It would not have resolved the problem of the fixed annual demand, however, and was anathema to the Survey Department, who saw the region's poverty as proof of its potential to improve, from which the state must be able to benefit. It was this attitude that campaigners against the Bombay Settlement blamed for *ryots'* difficulties. The *Poona Sarvajanik Sabha* declared 'We think that Government, for once and all, should give up any hope of increasing its land revenue in the arid or dry Deccan districts.'[111] They called for a *zemindari* system in Bombay, in which they doubtless saw themselves as the *zemindars*. Dutt's optimistic assertion, too, that a landlord system would have allowed indigenous capital to accrue to the benefit of the whole agricultural community and prevented subsequent famines,[112] reflected the conflict between Indian elites and the colonial state. The experience of tenants in Bengal suggests that the plan may not have been so beneficial for poor Deccan farmers. The relative influence and varying agendas of those proposing alternative revenue suggestions was significant. Most were raised at lower levels of the colonial hierarchy, by Indian campaigners, or at best in discussion documents like Wedderburn's, Baden-Powell's later review or internal memoranda of a general nature in the distant imperial Revenue and Agriculture Department.[113] Moreover, while even the Survey Department acknowledged that revenue rigidity added to the natural difficulties of the *ryots*, even British opponents of *ryotwari* also saw it as efficient and not to blame *per se* for poverty – making it consistent with colonial notions of good administration. The Government of Bombay could therefore dismiss all rival ideas, declaring with finality in 1884, 'it is not our intention either to abandon any of the principles of Land Revenue Settlement approved by this Government in past years or to adopt any new ones'.[114]

It had acknowledged in 1874, however, 'the need for moderation, and for assimilating the results of the revisions which were effected when prices were comparatively high, to those which will henceforth be made under altered circumstances'.[115] It might have been expected, then, that revisions of assessment in Ahmednagar, most of which followed the famine as well as declining cultivation and profits, would be conservative. Wingate himself argued that if the state had the aim 'of fostering and developing the resources of a country so situated, the demands of Government cannot be too much reduced'.[116] In calculating the new rates, however, the Survey Department remained determined to charge for benefits from the whole 30-year period since the first settlement, as well as for anticipated future development, on a selective, optimistic basis. Thus factors like access to railways, the benefits of which had yet to be proved as they were still being constructed, were accounted for, whereas the contraction of credit following the Deccan Agriculturists' Relief Act was not seen as sufficient grounds to sacrifice a full 30 years of potential revenue. It was pointed out that Wingate had set his rates at a time of even more severe economic depression, when transport facilities were virtually unusable. Yet his own calculations had erred on the side of caution, in order to ensure *ryots*' solvency as far as possible.

It was argued, however, that Wingate's first survey measurements had also been careless, creating a lack of uniformity that undermined the scientific basis of the system. In particular, Francis, who initiated the revision survey, repeatedly claimed that the original survey had been over-generous in Ahmednagar, where much of the soil was more productive than it looked.[117] Captain Davidson, who had surveyed the district for Wingate, had used a somewhat different approach to that in the rest of the presidency.[118] The determination to correct such perceived errors by full re-measurement was more important to Francis than the poverty of the district, which he blamed, ironically, on falling produce prices as opposed to the revised assessment.[119] Besides, his increases were too small to make as much difference to *ryots* as their limitation would to the treasury. This was firmly rejected by Havelock, who insisted, 'ten rupees for the field may appear a small sum to rich officials, but it was evidently a large one to the poor wretch who appealed to me'.[120]

The revision settlement was a daunting enough prospect for Ahmednagar *ryots* before Francis' decision to conduct the entire survey again first, instead of measuring only to take account of extension of cultivation. This was invasive and created increased uncertainty for holders, whose assessments could rocket if their fields were re-categorised up the scale. The re-measurement and classification of every field in the district was also time-consuming and expensive. While it was claimed that this would make the rates permanently fairer *vis a vis* each other, the aim was primarily to recoup revenue which was deemed to have been unnecessarily wasted by Wingate's laxity. As part of the reason for this was perceived to have been corruption and inefficiency on the part of native surveyors, the new survey was to be done only by Europeans, further increasing its cost. When called to justify such a large drain on resources at a difficult time, the Bombay Government convinced that of India that 'the annual increase in the land revenue, arising from the enhanced assessments made at re-settlements, is so large that no question can exist as to the financial advantage of continuing and completing the revision of the settlement'.[121] In the Northern Division, including Ahmednagar, Sholapur, Nasik and parts of Poona, the total cost of revision up to 1879 had been Rs 1,235,909, while the consequent annual increase in revenue amounted to Rs 667,043, thus recouping the cost within two years at the direct expense of *ryots*.[122]

Even given its return, the huge cost of the survey was a concern at all levels of government. When Francis' successor, Colonel Anderson, retired as Survey Commissioner in 1880 and the Government of Bombay applied for sanction for Stewart to replace him, Ravenscroft, the Chief Secretary, appended a dissenting note against the continued outlay.[123] Hartington went further in attacking the entire revision survey, suggesting that it caused 'harassment ... trouble, anxiety and expense' for the peasantry, that anyway 'proportionate equality of assessment is not always practicable under the law' and that cutting down on field surveys would save both effort and delay in setting revised rates. He only sanctioned the retention of the office of Survey and Settlement Commissioner 'on the understanding that it is not to be considered a permanent appointment', and that re-measurement or re-classification would only be carried out where it could be demonstrated in advance that it was needed.[124] After taking office, Stewart protested that judging which areas would need remeasuring was a 'stupendous task', impossible without his staff inspecting them anyway, and 'a very great responsibility ... upon a single officer'.[125] The Bombay Government consequently persuaded that of India that all areas should be re-measured once and for all, including those not due for revision for several years. The Survey Department could then be disbanded, with responsibility for future assessments left to the Revenue Department, assisted statistically by the new Department of Agriculture. This was at least an acknowledgement of the need to cut down bureaucracy. Granting jurisdiction over rates to revenue officers who spent their time in the district was also an important shift in the long term, judging by the frequency with which Ahmednagar collectors raised objections to the Survey Department's particular revisions. In the short term, however, it did nothing to reduce the expense of the land revenue system.

Revenue revisions in Ahmednagar district

In 1874, Collector Boswell wrote a special report on the condition of cultivators in Ahmednagar and called for an enquiry into the possible causes of their poverty before any revenue revisions were introduced in the district.[126] This was already late, with the first coming in 41 villages of Kopargaon *taluka* in 1874–75,[127] a precedent which left Boswell expressing 'some concern as to the future'.[128] 'It must be acknowledged', he declared, 'that it is one thing to impose higher though still fair rates, on a thriving and contented population; and quite another thing to enhance by ever so little the rates paid by people already in the extreme of poverty and almost goaded to desperation'.[129] Had they happened to commence in poorer *talukas* like Shrigonda or Sheogaon instead of relatively safer Kopargaon, he claimed, the results of what he called the government's 'experiment' would have been 'more disastrous'. As it was, he made dire warnings as to the outcome if nothing was done to improve people's conditions. The rise was 'a step of considerable hazard, for who could say that even this moderate increase might not prove as the last straw on the camel's back, and so complete the ruin of the ryots'.[130] While this was not necessarily intended to be a warning of famine, it can be taken as such in retrospect.

These comments were reprinted as an appendix to the Deccan Riots Commission Report, which as has been seen agreed that revenue rises were proving too much for the peasantry.[131] According to the Sub-Judge of Ahmednagar, the riots themselves had been in response to fears that peasants would soon face revisions as high as those already imposed in Poona.[132] A year later, there had still been no response to Boswell's arguments from the Bombay Government, causing him to repeat his prescient admonition that *ryots* could only survive with state assistance:

> Discontent to a great extent prevails among the agricultural population and though they have received a lesson as to the inutility of open rioting, yet if nothing is done to ameliorate their condition & remove the evils they are suffering from it cannot be doubted that their discontent will increase and will speedily again manifest itself, next time probably in some form more difficult to deal with.[133]

Pedder, a former survey officer, also argued that even a theoretically justified revenue hike of 50–100 per cent, such as those Francis had introduced into several villages, 'must still produce considerable inconvenience and perhaps hardship. For it not only diminishes suddenly the amount which a ryot has become accustomed to spend on his living, in other words, his "standard of comfort", but it causes a great change for the worse in his relationship with his Sowkar'.[134] The Deccan Riots Commission too asserted 'There can be no doubt as to the effect of such enhancement upon the indebted cultivator. By diminishing his profits it renders him less able to repay his present debts, and it also renders him less able to borrow.'[135]

Thus, familiar arguments about the land revenue demand were quickly rehearsed when revisions started in Ahmednagar, and a familiar stand off emerged between local revenue officers and the Survey Department, exacerbated by a lack of dialogue between them. Part of Boswell and his successors' frustration was that survey officers conducted revisions alone, 'without any authoritative intervention on the part of the Collectors', who were only able to make general criticisms on the completed report when rates were sent for sanction.[136] However, the biggest row between the two departments, in 1874, focused on Revenue Commissioner Havelock. It related to Wingate's land classification, which the Survey Department believed had been too generous when classing many fields as *pot kharab*, unarable, making them free of all assessment bar a nominal grazing fee. Many such holdings had subsequently come under cultivation. Revenue officers, on the other hand, argued that this was such poor land that it was only viable because of the absence of revenue, and that its re-classification would prove unbearable as well as amounting to an infinite rise.[137] While it was argued that extension of cultivation proved the previous survey had been low, Sumit Guha's demonstration that such new cultivation made peasants of those without the resources to farm, on the worst possible land, means there was little to distinguish such holders from landless labourers.[138] It is likely that they also worked for others and were similarly vulnerable to famine. Re-categorisation of *pot kharab* lands as cultivated had the added advantage of softening the statistical impact of declining cultivation, further deflecting charges of over-taxation. Yields per cultivated acre would have fallen correspondingly, but so would revenue per acre.

The *pot kharab* debate was entangled with that of self-improvement of holdings. Under the 1871 Land Improvement Act, revenue could not be raised in the Deccan on account of investment in land by the holder.[139] This valuable concession, which did not apply in more fertile districts, was designed in particular to encourage *ryots* to build wells (if necessary by taking *takavi* loans), without fear of incurring the much higher *bagayet* rates for irrigated soil.[140] Havelock suggested that this rule ought to apply to *pot kharab* lands, many of which had only been made cultivable after considerable exertions of labour to clear scrub and stones and break up rock-like soil before ploughing.[141] While Francis continued to deny that all uncultivated land had actually been uncultivable when Wingate classified it, his main argument was that the rule did not refer to investments of labour, only of capital.[142] Thus, in keeping with general agrarian policy, benefits would be made available to potentially capitalist farmers that were specifically denied to the poor, and therefore served no protective function.

More significantly, according to Havelock, the creation of a new soil classification for formerly *pot kharab* land had pushed the previously lowest classes of land up into new bands which did not reflect their own improvement. As a result, the lower and middle ranking soils were hiked by a greater degree than the best, notwithstanding Francis' greater imposition on the highest bands, because they had been categorised higher as well as revised within their new category.[143] Havelock then challenged the supposed scientific nature of a system of re-classification which started by putting lower soils at a disadvantage, removing the basis of the success of Wingate's settlement, which had made the greatest reductions from Pringle's rates in the lowest classes.[144] Claims to greater accuracy of measurement missed the point, he insisted, 'from the point of view at which a cultivator or a revenue officer, anxious for the well-being of the cultivators, would regard it'.[145] Francis refuted both charges strongly, insisting that only survey officers were qualified to judge classifications: 'Now as Mr Havelock, it may be assumed, has not the practical knowledge of the working of the system ... it is temerity, I think, on his part to say that certain land which he has not been able to identify has been wrongly classed.'[146] His lack of sympathy for *ryots*' views was not, however, denied: 'I can well understand the increase caused by rectification of mistakes in the former classification, being distasteful to the cultivator, but his opinion is not worth much in a case where self interest is so deeply concerned.'[147] While this justification ignored the state's own interest, Francis responded to the suggestion of unfairness by arguing that much *pot kharab* land had anyway been taken up by illegal encroachment, on which it would even have been fair to backdate the demand.[148]

The Government of Bombay, however, opposed its survey commissioner, ordering that all land previously unassessed should remain so.[149] This then provoked an angry response from the Government of India. Though the Viceroy, Lord Northbrook, admitted that the measure's 'substance is, I think, wise',[150] they objected to the way it had been decided, without consultation, by a government resolution, which was seen as a deliberate attempt to undermine their supreme authority.[151] The ruling, they averred,

> affects the revenue system of nearly the whole empire, and as it is novel and opposed to pre-existing practice alike in Bombay and elsewhere, His Excellency the Governor in Council will doubtless, on re-consideration admit that its adoption is a matter of grave imperial importance, demanding the fullest consideration by the Government of India as well as by the Government of Bombay.[152]

Echoing Francis' response to Havelock's charges, the Government of India thus showed more concern for the limits to its antagonist's authority than for the force of their argument. They ordered that re-classification of *pot kharab* land must be permitted.

Following this verdict, large amounts of formerly unarable land were re-classified in the process of Ahmednagar revisions, until the Secretary of

State raised the issue again in 1883. By now, the central government opposed the levy of revenue on previously *pot kharab* holdings, admitting their fault in rejecting Governor Wodehouse's edict, 'of which the main object was to remove the annoyance of the re-measurement and re-classification of the land'.[153] It was therefore pleased when the Government of Bombay, which they expected to defend existing policy, offered the compromise suggestion that Wingate's original survey methods had been curiously improved by 1854. Thus in those villages whose 30-year revisions were still to be done, *pot kharab* land could be left as it stood.[154] For Ahmednagar peasants, though, the results of political rivalries way beyond their world had already proved disastrous. Between the rejection of the *pot kharab* concession from above in 1874 and its specifically non-retrospective acceptance from below a decade later, almost the whole district had been re-surveyed, and the majority of revenue revisions implemented.

It is nonetheless interesting that both governments reversed their position on the issue in this period, their sole consistency coming in opposing each other's views. This reflected a broader turnaround on revenue revision questions. In 1874, the Bombay Government responded to concern about early revenue revisions in the Deccan by imposing a ceiling of 33 per cent for any area or group of villages being revised simultaneously, which was well below the average revision rate up till then. Further, if any single village's assessment was raised by over 66 per cent, or any particular holding over 100 per cent, the Survey Department was obliged to provide a special explanation in its settlement report.[155] This was a meaningful limit, but the Government of India protested that concessions for a poor region risked setting a precedent that secure areas could then demand. The Revenue Member of the Viceroy's Council noted 'that in the district of Ahmednagar, stated to be more depressed than any other, the rates were settled 25 years ago, and are exceptionally light'.[156] Thus hardship under a low revenue demand did not justify an attempt to keep it low. Following this, the imperial government requested details from all provinces of the settlements now in progress, including their length and anticipated levels.[157] This reinforced their overall control of the process, and encouraged the Bombay Survey Department to continue to maximise the opportunity for revenue increases. Only after hardship had turned into a politically embarrassing and spectacular famine event did the central government make Ahmednagar their concern, as revenue revisions continued in the district.

The first to object to excessive post-famine revisions was Cranbrook, who was unhappy to see Bombay revenue revisions averaging 44.5 per cent for 1877–78, 'almost entirely in the four Deccan districts which have suffered most from the recent famine'.[158] This included new rates in 56 villages of Kopargaon *taluka* that had been postponed from 1876–77, although the second year of the famine was actually worse in the north of the district. In 1879, the Government of India suggested that all revenue revisions in the Deccan should be postponed to aid recovery not only from the famine but also the initial impact of the Deccan Agriculturists' Relief Act. When rebuffed, they asked the Secretary of State to impose a five-year moratorium.[159] This would have only applied in Ahmednagar, as the revisions in Poona and Sholapur had been completed, and Satara was not due for revision within that time. The Government of Bombay now echoed the central government's earlier position, insisting:

> The increase cannot be postponed without admitting that the weight of the land revenue demand is one cause of the poverty and indebtedness which prevail amongst a portion of the cultivating classes, and is heavier than the profits derived from the tilling of the soil can bear. But this is not the fact.[160]

Nor was the Government of India's shift lost on them: 'But for the two recent years of famine and scarcity no question would have arisen regarding any grant of relief of land assessment to the ryots.'[161] Yet their own position had altered equally radically. They now recalled Francis' views when he re-assessed *pot kharab* land, arguing that the cost to the exchequer of postponement would be greater than the trifling benefits felt by *ryots*. In 1874 they had rejected this logic as insensitive to peasant hardship. By further suggesting that the effects of the famine had not been so great as the central government assumed, the Bombay Government came close to admitting that it had not been far removed from the normal condition of the district, whether taxation was the cause or not.[162]

Faced with the complex and emotive issue of a possible moratorium, the new Secretary of State, Hartington, concluded in September 1880 that the question of jurisdiction was paramount: 'the measures now recommended for my sanction are strongly opposed to the opinion of the Local Government, whom it is proposed to overrule on a matter of internal administration regarding which it is admitted that they have the best means of forming a correct judgement'.[163] The central government protested, but in private Revenue Member C. U. Aitchison admitted that what irked him most was lost pride: 'I think that in this matter the Secretary of State has treated the Government of India with scant consideration.'[164] Power mattered more than peasants. Hartington had not had his last word on the matter yet, however. Reviewing the latest revisions two months later, he vigorously condemned the rates set in 99 villages of Rahuri *taluka* and 82 of Sangamner.[165] Not only were the respective rises of 40 and 32 per cent inappropriate, he backed Havelock's old arguments against new categorisations of land which pushed revisions in real terms over 100 per cent per village, and more on some individual holdings. The claim that this was a product of improved measurement was dismissed in a wholehearted defence of the accuracy, uniformity and authority with which Wingate had initially settled these regions. He was therefore now more persuaded by the case for temporary relief. Instead of postponing revision settlements entirely, however, he ordered that they be restricted to 20 per cent per *taluka* or village cluster for five years, not only in the Deccan but throughout the famine districts.

This created considerable confusion. The Survey Department still had free rein to calculate rates, but temporary remissions were to be granted on the part of the increase over 20 per cent. Given the variety of different bands and rates which made up each settlement average, such calculations would have required almost as much paperwork as the original revisions themselves. In the end, new revision rates were calculated but deferred entirely, with one to three *annas* added to each holding as an arbitrary interim increase.[166] In already revised areas such as Kopargaon, the survey commissioner placed selected villages in lower bands and then deducted two *annas* from each band's maximum rate. Even these calculations took so long to consider that for the first two years after Hartington's edict, no reductions were made to the levy at all. Once they had finally been calculated, these years' remissions were granted as a credit against the third year's demand.[167] This bureaucratic delay diluted a weak measure to a point where it served little use at all. More importantly, it provided no check on revision levels. Even during the partial postponement, new revision rates were introduced as planned in 1881 to 22 villages in Kopargaon, 33 in Rahuri and 50 in Sangamner.[168] Collector King particularly questioned a 43.9 per cent rise in villages near Juala Bhuleshvar in Sangamner which were prone to water scarcity and 'hardly so uniformly productive as to bear this increase', and one of 81.8 per cent in the Rahuri village of Chikalthan.[169] By 1882, concern was far wider spread, with the Government of India being joined by Hartington and Robertson in condemning the revision rates in Newasa and Rahuri *talukas*, and indeed the Survey Department's general method.[170] Ironically this widespread outcry greeted the first revision settlement by Stewart, the recent Collector of Ahmednagar.

High level concern for the fate of poor peasants after the famine thus did little more to help them in the face of heavy revenue hikes in

Ahmednagar than Boswell's advocacy or Wodehouse's even-handedness had done before. When called to consider extending Hartington's reductions in 1883, the Bombay Government acknowledged that 'It is probable that, owing to the depressed condition of many of the cultivators remissions will be necessary for several years.' They also reduced the controversial hikes in Rahuri from an average of 40 to 31.1 per cent.[171] However, despite the run of poor seasons that had followed the cap on revisions, they opted to introduce the new rates in full the following year, effectively adding another hike.

It is striking that the Government of Bombay was more concerned about the effects of land revenue revisions before the famine than after it. They were never excessively generous, but were at least prepared to look into the question of agrarian poverty and to try to mitigate the harshest effects of the hikes. When the state's interests were under pressure during and after the crisis, however, balanced enquiries took a back seat. Revenue losses and relief expenditure, as well as changes in personnel, adversely affected their attitude towards peasants, as is highlighted in the following section. Conversely, the Government of India had little interest in the well-being of under-capitalised Deccan *ryots* before 1876 and refused to allow concessions to them, but the shock of the famine event turned their attention to the long-term problem of the revenue burden. It was a significant irony that the central government's power to overrule, which had thwarted Wodehouse, failed later when they in turn sought to redress the impact of the revenue. As the two arms of state circled around each other on revenue policy, the power of veto seems to have been retained by whichever took the harsher line for the *ryots*.

Land revenue collection during and after the famine crisis

Under the Bombay land revenue system as it stood before the famine, remissions could be granted in individual cases of extreme poverty, desertion or accidents, such as fires, floods or extreme cases of crop failure. Discretion was left to local collectors, subject to government ratification, and this was retained in seasons of scarcity, such as that of 1871–72 in Ahmednagar, so that the total of individual remissions could be considerable. Thus, before the Famine Codes were written, local remissions were the main form of concession to scarcity conditions, including those confined to a single district, which were not definable as famine. Revenue remission had been the first response to famine or scarcity under *Maratha* rule, but its use under the British was on an *ad hoc* basis and never recognised in the official Revenue Codes. Formal revenue suspension – permission to pay revenue the following year – was rare, although outstanding balances were recorded without extra pressure to pay being exerted in some poor seasons. Table 4.3 shows the levels of remissions and outstanding balances in Ahmednagar district from 1867–68 to 1882–83.

The total demand column reflects the steady decline in cultivation prior to the famine, with almost 50,000 fewer rupees expected in 1874–75 than 1870–71. Thereafter, the demand increased as revised rates were introduced, with the exception of 1879–80, when charges on holdings abandoned during the famine were finally given up as 'dead loss'.[172] That the demand at the end of the period was so close to that in 1867 reinforces the impression that revenue revisions were intended to recoup ongoing losses from declining cultivation as much as to claim a share of profit from its earlier expansion. The Bombay Government's reactions to Annual Administration Reports also suggest a constant pressure on district revenue officers not to let their pessimism about the condition of the peasantry be reflected in their levels of collection, before or after the famine. While the column showing amounts collected each

Table 4.3 Land revenue remission, suspension and collection (rupees), Ahmednagar district, 1867–83

Year	Total revenue demand	Remitted	Levied	Left outstanding	Collected
1867–68	1,385,394	1,106	1,384,288	—	1,384,288
1868–69	1,382,632	39,802	1,342,830	1,523	1,341,307
1869–70	1,368,693	4,937	1,363,756	442	1,363,314
1870–71	1,376,607	459	1,376,148	156	1,375,992
1871–72	1,355,334	46,898	1,308,436	49,832	1,258,604
1872–73	1,358,609	2,087	1,356,522	2,961	1,353,561
1873–74	1,344,246	833	1,343,413	3,025	1,340,388
1874–75	1,328,622	1,166	1,327,456	74	1,327,382
1875–76	1,339,756	16,210	1,323,546	2,441	1,321,105
1876–77	1,355,478	131,741	1,223,737	351,261	872,476
1877–78	1,354,602	—	1,354,602	366,193	988,409
1878–79	1,362,597	—	1,362,597	176,603	1,185,994
1879–80	1,309,595	92	1,309,503	77,093	1,232,410
1880–81	1,366,154	—	1,366,154	126,514	1,239,640
1881–82	1,379,330	220,884	1,158,446	181,102	977,344
1882–83	1,395,498	164,842	1,230,656	120,104	1,110,552

Source: Ahmadnagar Gazetteer, p. 555.

year adequately reflects the scarcity years of 1871–72, 1876–78 and 1881–82, it is striking that their reduction did not come through remission between 1877 and 1881. This was the result of Temple's controversial attempt to limit the cost of the famine to the state. The corresponding rise in outstanding balances included both formal suspensions, granted instead of remissions in those years, and unpaid levies, the collection of which was prioritised over the annual demand in the following year.

When the extent of the 1876 crop failure became clear, the usual procedure for consideration of remission claims commenced. At first, the Government of Bombay encouraged remission in badly affected areas on more generous terms than usual, announcing that 'in villages where there is nothing but kharif, and where that crop has wholly failed, entire remissions may be granted'. Where the crop was below eight *annas*, 'the first instalment may be wholly foregone'.[173] Peasants with harvests over that arbitrary figure should pay in full. Thus, within limits, revenue remission was to be a significant component of the response to the famine in Bombay: 'It is the wish of Government to show the people the utmost consideration, but at the same time not needlessly to forego public revenue.'[174] However, the Government of India, keen to take control of famine policy as well as revenue matters under Strachey, requested in January 1877 that 'no absolute remissions or promises of remission may be made for the present', with revenue only suspended in cases of need.[175] This would enable the state to recover its losses if the subsequent season proved good enough for an effective double levy. The Government of Bombay opposed the new policy, declaring, 'this Government is of opinion there should be entire remission of land revenue to such extent as may be necessary'.[176] Many remissions had already been granted with the central government's knowledge under long-standing local procedure.[177] When the latter asked 'whether His Excellency the Governor in Council anticipates that any evils will arise from suspending, instead of absolutely remitting, land revenue',[178] Bombay responded that the change of policy would save the state little and cause considerable harm:

> It is really immaterial whether the collections are suspended or absolutely remitted. They never will or can be collected, and if they are allowed to hang over the heads of the impoverished and desponding tenants, who in the majority of cases live from hand to mouth on the proceeds of a few fields and have no capital to fall back on, the effect will be that the debtors will never resolutely try to grapple with their misfortunes, and the village bankers will naturally be disinclined to deal, except on most usurious terms, with cultivators,

whose sole security can at any time be taken possession of in payment of outstanding balances of land revenue.[179]

The Government of India did not insist, so long as 'Collectors are again warned against too free remissions'.[180] Strachey, however, had already ensured an eventual change in the local government's opinions by appointing Temple as the new governor. Before this was even confirmed, Temple wrote to Lord Salisbury, promising that 'fiscal interests, revenue + other, shall be guarded against undue sacrifice', especially during the famine. To this end, he already planned to suspend rather than remit revenue.[181] Before taking up his post in May 1877, he also unsuccessfully attempted to persuade the Government of Madras, as famine envoy, to abandon its wholesale remission of land revenue, which Strachey later preposterously blamed for India's entire revenue shortfall in the 1870s.[182]

During this time, Temple also contradicted his future colleagues' predictions of revenue losses from the famine. At the Government of India's request, collectors in the 11 famine-affected districts of Bombay estimated the extent of likely remission for both the current year and 1877–78. As this did not assume the second monsoon failure which eventually occurred, this shows that they initially anticipated losing further revenue while *ryots* recovered from the crisis. Their figures were then revised upwards by Rogers of the Bombay Revenue Department for budget purposes. Both are shown in Table 4.4.

After a brief visit to Bombay in January, however, Temple told Calcutta that he estimated a total revenue loss in the presidency of only 50 *lakhs* (5 million) rupees over both years. Stung, the Bombay Government retorted, 'in the absence of explanation as to the method by which Sir Richard Temple arrives at his diminished total of 50 lakhs, this

Table 4.4 Anticipated land revenue losses (rupees) due to the 1876–78 famine, Bombay Presidency

Estimate	1876–77	1877–78	Total
Collector of Ahmednagar (Jacomb)	463,000	250,000	713,000
Rogers' revision for Ahmednagar	513,000	250,000	763,000
Collectors of all famine districts	5,150,000	1,915,000	7,065,000
Rogers' revision for all famine districts	5,494,000	2,370,000	7,864,000

Source: GOB to GOI, No. 606, 30 January 1877; MSA, GOB, RD (Famine Branch) Vol. 58, pp. 60–1.

Government would prefer abiding by their previous opinion', which had been 'carefully prepared'.[183]

Jacomb was quick to submit *jamabandi* settlements for Ahmednagar, incorporating his revenue remission proposals. By the end of May 1877, over half the remissions sanctioned in Bombay Presidency had been in the district, including parts of Jamkhed, Karjat, Kopargaon, Nagar, Parner, Rahuri and Sangamner *talukas*.[184] Temple's accession then coincided with the receipt of proposals from Woodburn for the largest and most severely famine-affected *taluka*, Shrigonda, containing remissions of 72,108 rupees, further suspension of 13,806 and current revenue collections of a mere 29,273. Now the Bombay Government responded that all remission or suspension decisions would be deferred till after the monsoon had broken.[185] When it again failed, they still allowed no more revenue to be remitted, pronouncing that 'collection should, for the present, be suspended. But none of it must be written off as irrecoverable. The whole must be treated as outstanding balances, and so entered in the accounts'.[186] The refusal to remit revenue reflected the view that famine was a short-term natural disaster, without causes or effects relating to *ryots*' livelihoods. It therefore exacerbated the chronic problems it sought to ignore. The knowledge that extra revenue would be demanded soon after the famine made long-term prospects so bleak that many cultivators permanently abandoned their land. As the Government of Bombay had pointed out before Temple's arrival, this process was accelerated because many creditors refused to lend for revenue demands they had not expected, and in some cases evicted debtors immediately, before the state could do so for famine arrears. The non-remission policy therefore did even more to harm peasants' chances of maintaining their livelihoods through the famine crisis than the impending revenue hikes. Nonetheless, it remained in place in Ahmednagar until 1881–82.

In his final minute on the famine, Temple professed satisfaction that Rs 2.7 million of land revenue remained suspended and only 216,000 rupees remitted (of which 131,741 had been in Ahmednagar before he arrived), using these figures to demonstrate that proper regard had been given to economy during the campaign.[187] His similar meanness over famine expenditure allowed him to claim that the total famine cost to the Bombay Government was just Rs 11.6 million, as is seen in Chapter 5. In this context, non-remission represented a significant saving, but only on the assumption that all suspended revenue would ultimately be recouped. The attempt to do so after the famine provoked condemnation from the Secretary of State. In 1879, Cranbrook strongly criticised the

Government of Bombay for keeping famine revenue arrears suspended, when they had only succeeded in collecting 550,000 of the 3,250,000 rupees originally postponed in the Presidency by the end of 1877–78: 'It seems probable that it will take several years longer to realize the famine arrears of 1876–77, even with a rigour of collection not unlikely to produce evil effects as regards the condition and contentment of the agricultural classes.'[188] He therefore ordered that all outstanding famine balances should either be collected in the current season or deemed irrecoverable, leading to 1,548,170 rupees being written off in 1880.

Thus the final figure for remissions resulting directly from the famine was eight times higher than Temple had boasted in his minute, and the total cost of the famine to the Government of Bombay was 12 per cent higher. He had nonetheless succeeded, by deferring some of its costs to the state, in creating an impression of economy that remained on record. Similarly, keeping suspended revenue demands in the accounts meant that large areas of land were still regarded as held after being abandoned by *ryots* during the famine, until such arrears were written off in 1879–80. This not only minimised the apparent effects of the famine crisis, during which only 17,000 acres were recorded as having gone out of cultivation in Ahmednagar, but also to render statistics meaningless subsequently. As seen in Chapter 1, recognising so many departures belatedly disguised new resignations. One effect of the refusal to remit revenue during the crisis was, then, to make the Bombay Government's famine management look better than it was by diffusing its financial and human losses. In the process, peasants' hardship was also extended.

A series of further poor seasons in Ahmednagar made the collection of the annual demand without remissions difficult in the aftermath of the famine, and the retrieval of suspended famine balances still more so. District officers therefore continued to leave large outstanding balances each year. Even if the notion that one good season after the famine would remove all revenue shortfalls had been correct, in reality revenue arrears remained suspended over *ryots'* heads for several years. As the district Annual Administration Reports had to be produced in July, they could only record revenue collection up to the end of June, before it was completed. This meant that the figures were complicated, and rarely tallied from one year's report to the next. While a neater (but also non-tallying) retrospective picture was provided by the 1884 Ahmadnagar Gazetteer (see Table 4.3), it is interesting to explore in detail how collectors approached and reported the thankless task of recovering revenue in the aftermath of the famine.

In 1876–77, 816,555 rupees of the expected 1,152,853 had been collected by the cut-off date of 30 June, with a further 58,698 anticipated during July, leaving an outstanding balance of 277,657, in addition to the sanctioned remission of 131,741 rupees and 205,174 of formal suspensions.[189] By 1877–78, the previous years' suspensions had been revised down to 201,569.[190] Of the current year's demand of 1,352,460, only 944,787 rupees had been garnered by 30 June, with a minimum of 39,120 more expected. In the absence of remissions, this left a massive outstanding balance of Rs 368,533, prompting Collector Stewart to request permission to sell abandoned land for arrears immediately, without success. The high figure was also partly explained by the collection of 136,897 of the 1876–77 balances as a priority. According to Stewart's calculations, this left 229,803 rupees of the famine balances still to collect, a large proportion of which he declared to be 'absolutely irrecoverable'.[191]

The best season in the post-famine period in Ahmednagar was 1878–79, but it was not sufficient to eradicate such large arrears. Having taken over from Stewart shortly before the annual report was due, Collector King took the combination of the previous two years' balances to be Rs 578,620, of which a full 465,565 had been collected, leaving just 113,055 still on the books. Again, this had been prioritised over current collections, with 172,157 of balances left on the year.[192] Nonetheless, the recovery of so much of the state's famine losses was remarkable in the circumstances and can only have been at the expense of peasants' own recovery. The Government of Bombay admitted as much, while expressing satisfaction with Stewart's work: 'The amount of pressure required to realise the revenue was no doubt considerable, but it was less than in the previous year, and was judiciously and carefully exercised.'[193] This was scarcely true. Table 4.2 reveals that distraint of property increased fourfold in 1878–79 and sales of property and land raised over ten times more than in the previous year. Given the continuing bar on sales of abandoned land, this was from peasants attempting to continue cultivation. In a Freudian slip, King's clerk recorded levels of 'distress' and sale instead of distraint.[194]

Once more, 1879–80 produced large outstanding current balances of Rs 76,968. Of the 318,994 King recorded as left over from the previous three years, 182,524 were collected, leaving 136,470. At this point, all but 13,083 of the famine balances were finally written off as irrecoverable under Cranbrook's orders, amounting to a total loss of revenue of 123,387 rupees, slightly less than the remissions originally allowed in the district in 1877.[195] This was far lower than in Poona, Sholapur or Kaladgi, the districts classed alongside Ahmednagar as having been worst hit by the famine,[196] thanks to Stewart's vigorous collection of balances in the preceding year. Even then, a larger total of balances, 127,643, dating from after the famine and 1879–80 itself, was left on the books at the start of 1880–81. Thus, by always retrieving famine balances first, collectors had been able to minimise the write-off of revenue debts deriving directly from Temple's suspension policy, while still leaving Ahmednagar *ryots* with indirect arrears from the famine. While Rs 94,808 of these old balances were collected in 1880–81, and a further 857 remitted, this was a very poor season in which new balances of 130,644 were generated, mostly by deliberate suspension.[197] In the single *taluka* of Kopargaon, 52,850 rupees – almost a quarter of the revenue demand – were suspended in the year.[198] In the light of almost as large June outstanding balances in Sangamner (45,573) and Rahuri (44,322),[199] the granting of an official suspension in Kopargaon when remissions were still not to be granted suggested that local officers deemed that portion of the revenue demand to be unpayable.

This was confirmed when the entire amount was written off in the disastrous 1881–82 season. With remissions finally allowed again, their extent was remarkable, with between one-quarter and one-half of the demand being waived in northern *talukas* – more than in 1876–77 – although even then Rs 24,230 effectively of famine balances was not written off.[200] The final tally for crop failures and *ryot* insolvency in 1881–82 came to 303,796,[201] with a further Rs 120,685 suspended. Of old balances, 107,198 rupees – two-thirds – were remitted.[202] The Government of Bombay sought to limit the extent of these measures. Collector Elphinston recorded that having granted remissions early in the year, he was put under pressure to collect the remainder, despite his view that *ryots* would have difficulty meeting even the reduced demand.[203] His prediction was borne out, with the government expressing anger that a further 76,911 rupees of revenue beyond sanctioned suspensions were anticipated to remain uncollected during the year.[204] This was despite distress sufficient to necessitate the opening of relief works, five years after the famine event, significantly in the first *talukas* to have had their revenue rates increased.[205]

Even when the power to remit was restored to the collector, then, the Government of Bombay's pressure discouraged a generous response to peasant poverty. One outcome of this was that criteria for remission varied between *talukas*, which consequently received non-proportional relief. In 1881–82, Kopargaon suffered the worst rainfall, and received 149,883 rupees of remission on that account. Rahuri, which had *kharif* and *rabi* harvests categorised at four and five *annas* under the *annewari* system, compared to Kopargaon's two and four *annas*,²⁰⁶ only saw 64,126 remitted for rain failure.²⁰⁷ There were a further 30,983 of remissions granted in Rahuri for insolvency of *ryots*, but as this implied either the previous departure or the eviction of holders, it reflected, rather than alleviated, the difficulties of meeting the demand. The deputy collector responsible for both *talukas*, Anding, noted the difference, suggesting that Rahuri may have been treated less liberally despite similar need.²⁰⁸ This was further borne out by collections of unremitted revenue. By the end of June, only 33,992 had been received in Rahuri, leaving 90,349 outstanding.²⁰⁹ Although 34,936 in the *taluka* was then officially suspended, compared with 48,393 in Kopargaon, Rahuri still had outstanding balances of 54,145 rupees at the end of the year, comprising the majority of the 76,911 of uncollected revenue in the whole district that attracted the government's censure.²¹⁰

This was because general remissions on account of the poor season were only granted in Kopargaon. In Rahuri, as elsewhere, investigations of individual cases were required, under the old system – a monumental task. The *taluka* was seen as on the borderline, with many cases of hardship intermingled with those who could still survive, and was consequently subjected both to extra scrutiny and stricter judgement. Some villages were deemed to have such good crops that no inspections were made, leaving any individual difficulties unrecognised. Where crops had failed, however, distinctions were made, with some revenue relief given but other *ryots* deemed not to be 'so poverty-stricken as to entitle them to remissions' because they still had saleable assets, leading to increased distraints and sales.²¹¹ The Government of Bombay declared, 'it is demoralising to the people themselves to be allowed to evade the payment of dues which it is within their power to pay without difficulty'.²¹² Thus many local recommendations for remission were refused sanction. Despite the detailed inspections required, reports of inability to pay were rejected when they reached Bombay in the interests of simplistic discourse, and the supporting evidence of low levels of collection taken as a sign of weak local administration. The rigid obstacles to remission were further demonstrated in the chronically poor *taluka* of Shrigonda, which had also lost out to an accident of timing in 1877. Large old balances were left on the books again in 1882 because the Deputy Collector's proposals to remit them had been returned to him by government as lacking sufficient detail on the individuals concerned.²¹³

Even before the first remissions were allowed in Ahmednagar after the famine, Cranbrook put pressure on the Governments of both Bombay and India to establish liberal remission policies permanently. Bombay defended the suspension system, claiming in contradiction to the evidence of Table 4.3 that remissions had rarely proved necessary, except in extreme circumstances like the famine.[214] However, when later pressed to consider abandoning fixed revenue rates in vulnerable zones, they switched their argument again, claiming that remissions were sufficient, and had been 'freely resorted to, especially since 1877'.[215] Again, this was not true, at least in Ahmednagar. If remissions were given before 1881 in other districts, repeated resolutions instructing Ahmednagar collectors not to do so must have been intended to penalise the district for remitting so freely before the ban was first imposed in May 1877.[216] For its part, the central government took exception to Cranbrook's suggestion that Bombay had only abandoned remission 'in deference to the supposed views of your Government',[217] insisting 'Sir Richard Temple as Governor of Bombay very deliberately re-affirmed his own policy'.[218] In light of their telegram to Bombay on 16 January 1877, four months before Temple took up his post in Bombay, this was not true either.

In 1882, though, under new Viceroy Lord Ripon, they circulated a further resolution on the inelasticity of the revenue system, which had

> been forcibly pressed upon the attention of the Government of India by reports lately received on the agricultural condition of several parts of the country during the years immediately succeeding the famine, which show that the measures taken in 1877–78 for preventing the revenue from pressing too heavily on the people failed adequately to meet the difficulty, and that in many places serious and permanent mischief has been caused ... There can be no doubt that the rigid enforcement of the revenue demand, irrespective of the calamities of the seasons, was not part of the intention of the authors of the revenue system.[219]

However, remembering their lack of jurisdiction and Bombay's intransigent response to their request to postpone revisions, this circular was merely sent 'for information'. The Bombay Government was in no mood to change their land revenue system, nor to adopt generous remission rules for times of hardship, which would run the risk of encouraging *ryots* to claim '*as a right*' what should remain '*an act of grace*'.[220] Instead, in 1884, they laid down strict parameters even for postponement: 'When a Collector has clearly ascertained that an abnormal failure of the harvest, causing total or almost total destruction of the crops over a considerable area, is certain, he is authorized to suspend the collection of the next ensuing instalment of land revenue.'[221]

Conclusion

The land revenue demand was not necessarily the worst problem that Ahmednagar cultivators faced in this period. In several ways, though, it greatly exacerbated – and sometimes triggered – their existing difficulties. Inflexible collection, in timing and amounts, increased *ryot* vulnerability to the vagaries of the season and pushed them further into the hands of moneylenders. Large revenue increases at a time of low grain prices and crisis worsened the squeeze on credit and undermined their viability. The *ryotwari* system brought peasants and the state into contact, but also conflict over limited resources at a time of financial hardship for both. As a result, the Bombay Government's stance turned against *ryots'* revenue claims when their need was greatest. The extreme situation of famine prompted Temple's abandonment of standard remission policy, while the re-classification of *pot kharab* lands as taxable – albeit initially at the Government of India's behest – and the unwillingness to limit or delay revenue revisions hit the poorest holders hardest, exacerbating and extending the famine. A similar desire to protect the imperial coffers while trying to control the famine is seen in the Chapter 5.

The state was not, however, a monolith. There were always those who argued for various forms of lighter demand. Local revenue officers, for example, frequently criticised the actions of the Survey Department, with barely a revision settlement unchallenged to some degree. Radical ideas like seasonally floating revenue demands were seriously discussed. Colonial debates were lively on land revenue questions, and the appropriate extent of revisions or remissions in Ahmednagar was considered in Calcutta and London. But historical tensions between different departments or levels of government made them unwilling to compromise, each defending their corner more fervently than their views. Thus, no matter how impressive some officers' understanding of land revenue issues, the implementation of policy remained crude and simplistic. The central government's shock at the famine generated concern for struggling *ryots*, but only after the event, having refused to allow their sympathetic treatment immediately before it. The structure of control was as if designed to limit British sensitivity, especially when money was involved.

5
Peasants and Relief Labour

Introduction

Having examined the way in which two factors of production – land and capital – as well as markets put Ahmednagar peasants at a natural disadvantage by their low quality and unavailability, it now remains to focus on labour. Social relations of labour have already been discussed in the context of markets and credit. Relatively few *ryots* in the district were in a position to employ wage labourers. Rather, family members and peasants themselves often sought seasonal employment of their own to supplement their meagre incomes. Bullocks could also provide some labour power. Thus, this chapter will look at the peasant population itself. While it would be useful to examine the health and well-being of male and female household labour as a factor of production throughout the period, colonial data provides little information on peasant working patterns or labour conditions. The Annual Ahmednagar Administration Reports only recorded the scarcity of local industry, prevailing market rates for labour, and crudely collected population figures, which are examined at the end of this chapter.

More can be found on working conditions, health and mortality during the famine crisis, which the state sought to control by creating its own short-term labour opportunities as a form of relief. By the 1870s, it was well-established in colonial minds that Indian famines were not so much of food as of work. Jean Drèze and Amrita Rangasami have suggested that this oft-repeated mantra reflected some understanding of what Amartya Sen was to label entitlements, although the stage at which British administrators sought to protect the population's capacities and livelihoods could scarcely have been sufficient to satisfy Sen.[1] Coupled with the principle that government should not interfere with private trade in grain,² this meant that the 1876–78 relief campaign in Bombay Presidency revolved almost wholly around the creation and management of relief works. Some peasant households sought assistance from them, despite the British belief that *ryots* should not need to do so for survival. This chapter therefore focuses on colonial famine relief efforts. They were deliberately unrepresentative of the normal district labour market and not productive, being mostly infrastructural works. However, having emphasised the colonial sins of omission and commission that undermined peasant production, profits and security in the period, this study could not be complete without examination of the state's closest interaction with the Ahmednagar population.

While *ryots'* voices are as absent from the colonial archive as ever, internal discussions – and tensions – within the state were as plentiful on famine relief as any other issue in the history of British India. At times in the 1876–78 crisis, Ahmednagar's collectors and assistants, Revenue Commissioner Robertson and even Bombay Governors Wodehouse and Temple found themselves at odds with relief policies that were at once *ad hoc* and inflexibly designated from above. Though all attempted to argue within official channels, some of their observations can be read as statements of witness on behalf of the population, against the influence of dominant theories on policy-making. Analysis of debates and their outcomes is critical in understanding how ineffective and insensitive measures were adhered to in the face of famine suffering.³ The ways in which competing perceptions of the famine problem were played out within the colonial hierarchy at presidency and imperial levels have been discussed elsewhere, along with competing claims to expert understanding of the success of particular strategies.⁴ However, examination of junior officers' roles is also essential to evaluate the effectiveness of that hegemony. Native newspapers recorded that some British bureaucrats were 'so prepossessed by theories that they cannot see the plain truth of misery',⁵ while also reporting the transfer of two dissenting collectors – Hogg of Kaladgi and Grant of Sholapur – and the censure of Percival, Grant's successor, whose criticisms 'drew upon himself the displeasure of Government, without benefiting the poor famine-stricken ryots'.⁶ Though Ahmednagar officers' general loyalty was shown when Assistant Collector Spry was appointed in Hogg's place, their reports do reveal relief failures, disputes with superiors and some instinctive attempts to bend, if not break, narrow relief rules.

The key related questions to be asked of British famine policy in Ahmednagar are, thus, to what extent it reflected ongoing attitudes and practices – at any level – in relation to peasant labour before and after the crisis period of 1876–78; to what extent as a policy it was locally devised or ideologically driven; and how responsive it was to perceived ground realities, either in its creation or in its implementation.[7] Attention needs to be paid, therefore, to the criteria used in determining policy, disputes between different levels of government over policy, and native responses to both famine and relief (and British reactions to them). A district level focus on the practical processes of decision-making and relief implementation during the famine may shed new light on familiar issues – and throw up some others.[8] The Bombay Government's focus, after the famine, on the extent of mortality and the population's own responsibility for it, created fierce debates that were important but self-deluding. They deflected attention away from detailed assessment of the justice and effectiveness of particular aspects of the relief effort.

The chapter starts by considering colonial criteria for giving relief, expenditure constraints and its scale and character. Particular attention is paid to the application of the four tests (appearance, distance, labour and wage) upon which relief was conditional and the effects – including a labourers' 'strike' – of the reduced 'Temple Wage' mid-way through the famine campaign. Peasants' famine responses are discussed, with a focus on whether they applied for and took relief or emigrated, in the context of local implementation of relief decisions. The chapter will conclude by examining the extent of mortality, the efficiency and efficacy of local relief efforts and official reflections on the campaign – including assertions regarding the population and the drawing up of the Bombay Famine Code.

British relief criteria

As has been seen in Chapter 1, rainfall failures were common in Ahmednagar, but rarely attracted the sympathy of the state. Famine was narrowly defined in 1870, by the then collector, D'Oyly, as 'the *total* failure of crops over a *considerable tract* of country during any one year, or a *serious* failure of crops during a *series of years*, so that sufficient food does not exist in the land to support the inhabitants'.[9] This was somewhat contradicted less than a year later, when he opened 'famine Relief Works' after a *kharif* crop below half of the district average.[10] However the Government of Bombay did not record a famine and D'Oyly soon backtracked, predicting that 'distress will become famine' only if the *rabi* crop also failed.[11] This distinction between distress and famine was emphasised by the Government of India in 1877, when they announced:

> We say that human life shall be saved at any cost and at any effort; no man, woman or child shall die of starvation. Distress they must

often suffer; we cannot save them from that. We wish we could do more, but we must be content with saving life and preventing extreme suffering.[12]

Nonetheless, it is striking that they were still repeating the principles declared in Sir George Cooper's memorandum after the 1866 Orissa famine, in which he insisted that lives should be saved at any cost. The reality was rather different in Bombay Presidency. After the 1867 Famine Commission had advised presidencies to prepare anticipatory famine strategies, the Government of Bombay decided in 1871 that this 'should not be adopted in the territories under this Government, where natural and artificial advantages render the occurrence of a general famine almost impossible'.[13] Thus, with the market perceived as a safety net and famine defined so as to remit the government's responsibility to redress anything but massive crop failures, Bombay Presidency was declared not to be famine-prone, even as Ahmednagar underwent a significant scarcity.

While the scale of the 1876–78 famine, as shown in Table 5.1, was indeed unprecedented, this complacency had left no space for officers in Ahmednagar – or their superiors – to prepare for it. In the absence of advanced plans for works, the Bombay Government's initial relief efforts were confused, focusing from the start on the need to limit expenditure. The most clearly identifiable criteria for relief were therefore that it was reactive and late. Most relief programmes commenced in November 1876, by which time mortality had already been reported in native newspapers,[14] despite high prices and unemployment prevailing from the moment the *kharif* harvest failed in June.[15] Native newspapers argued unanswerably that Cooper's principles had now been abandoned

Table 5.1 Estimated scale of the 1876–78 famine, Bombay Presidency and Ahmednagar district

	Affected		Badly affected	
	Area (square miles)	Population	Area (square miles)	Population
Bombay Presidency[a]	64,063	10,037,000	38,677	5,830,000
Ahmednagar[b]	6,647	773,938	5,350	640,000

[a] Includes adjoining Native States for which data was available.
[b] Note that almost the entire district was thus seen as affected, and an above average proportion as badly affected.

Source: Famine Commission Report (FCR), 1878, part III, p. 166.

after proving too costly in Bihar in 1874.[16] Now providing relief frugally, regardless of human costs, replaced saving lives at any price.

Thus 1876 marked a departure in British famine policy in India. It also saw the creation of a benchmark which, via the 1880 Famine Commission Report and subsequent Famine Codes, established relief principles and paradigms of extraordinary longevity.[17] Retrospectively, it could therefore be argued that this famine afforded junior British administrators their last chance to dispute the adequacy of cost-cutting relief, or to challenge presuppositions about the famine process and the population. Largely, however, Ahmednagar officers restricted themselves to local disputes over relief management or finance, which can be read as critiques of the principles of relief set out in the Famine Commission Report,[18] but were not intended as ideological opposition. It is nonetheless significant that the Government of India sought to control famine policy in 1876–78. Fixed principles meant that relief policy necessarily failed to respond to events. Whereas the Orissa Famine Commissioners had argued that 'extreme awareness and liberal tendency' among local officers were more effective than 'mechanical obedience to a rule',[19] the Chairman of the 1880 Famine Commission, Sir Richard Strachey, was less inclined to trust his subordinates:

> it is my earnest hope that no temporary impulse of sympathy with present suffering, no selfish ... effort to escape at any cost the pain of witnessing it, may be permitted to stand in the way of that real benevolence which is founded on sound principles drawn by dispassionate intelligence from the lessons of experience.[20]

In September 1878, the Famine Commissioners and other experts, including Temple, met in Simla to agree these principles. Their discussions revolved around economy, effectively seeking to determine the minimum level of relief the state was obliged to give to maintain its legitimacy, with much emphasis on the efficacy of non-interference in private trade. Richard Strachey declared that the cost of famine, in monetary terms, was the only important criterion, suggesting that 'disturbing influences such as the pressure of public opinion in England or India' should be ignored.[21] Arguing that the 'calamity' of famine created an exaggerated impression of its importance, he asked at what level the pressure of taxation to finance prevention would make misery generally worse than occasional famines. Thus these critical discussions were not intended to devise an ideal famine policy; they took broader economic and political factors into account. A. C. Lyall acknowledged the validity of native newspapers' accusations on the Cooper memorandum by declaring, 'there are many principles which the state, or at any rate a free nation, ought to and does place higher than the question of saving everyone alive'.[22] It was not unreasonable to rank the importance of famine in this way, but peasants' perspectives on broader political and economic processes, or the links between taxation and famine, were also overlooked.

The determination of the 1880 Famine Commissioners to create a template for relief, in the form of a draft famine code by Famine Commission secretary Charles Elliott, was an improvement on their 1867 counterparts' hope that every British officer would familiarise himself with reports of previous famines.[23] However, it is evident that the commission did not intend to base its conclusions solely on the experience of the famines (1874 as well as 1876–78) it was set up to investigate. It is no coincidence that similar principles of fiscal stringency in providing the 'necessary evil' of famine relief had long been propounded by Richard Strachey's brother.[24] As Collector of Moradabad, John Strachey had written an influential minute on the subject, which he implemented during the North West Provinces famine of 1861. His time as Finance Member was also dominated by attempts to control provincial governments' budgets, which came to a head during the famine period.

Many local governments – particularly Bombay's – had indeed been profligate in the recent past, and a further imperative for cost-cutting was created by the 1873 collapse in the value of silver, to which the rupee was pegged, triggering depression in the Indian and global economies. As has been seen in previous chapters, this phenomenon reduced agricultural profits, the availability of credit and household stocks of gold jewellery, which would have aided survival strategies, particularly for women. Its effects on public spending were perhaps most significant of all. Land revenue and other receipts declined and the relative cost of foreign rule, involving payments in sterling, still linked to gold, rocketed. As the famine started, the Government of India ordered every imperial and provincial department to 'take prompt measures to stop all outlay of public money which is not absolutely necessary, or to which the Government is not committed'.[25] Further, Secretary of State Salisbury declared that presidencies should now finance relief from their own budgets, rather than being supported by the central government, as had happened in Bihar in 1874.[26] Retrenchment at this time was particularly unfortunate in the light of the visible expenditure that went ahead on both the Delhi *Durbar* (court celebration) and a purposeless war in Afghanistan that came to be known as 'Lytton's Folly'. Logic was not the Viceroy's strongest suit anyway, judging by his pronouncement that

'I am profoundly convinced that every rupee superfluously spent on famine relief only aggravates the evil effects of famine and that in all such cases waste of money involves waste of life.'[27]

It cannot have been difficult, then, for John Strachey to persuade Lytton to appoint Temple, first as Famine Envoy to Madras and then as Governor of Bombay. Temple's instructions, publicised in the *Gazette of India* in January 1877, were an explicit forerunner of the commission's views:

> While it is the desire of the Government of India that every effort should be made, so far as the resources of the state admit, for the prevention of deaths from famine, it is essential in the present state of the finances that the most severe economy should be practised.[28]

In turn, Temple told Salisbury of his approval: 'The Govt of India *must* have a dominant voice in determining the policy to be adopted if famine breaks out, otherwise the manner in which the two Presidencies, Madras + Bombay *began* to deal with the crisis of 1876–77 might lead to a dangerous degree of imperial embarrassment.'[29] In Bombay, at least, Temple succeeded in imposing fiscal stringency on relief management. The Bombay Government's famine policies were otherwise characterised by trial and error, which badly needed dispassionate evaluation of the type Richard Strachey applauded in Simla. The 1898 Famine Commission reported that since all aspects of famine policy in the 1870s had been 'virtually experiments in relief', the 1880 Commission had had to formulate policy from scratch. It was willing, however, to adopt Bombay's *ad hoc* ideas where they succeeded in limiting expenditure, even if they were demonstrable failures in relief terms.

Relief costs

While economy was accepted within the colonial hierarchy to be the top priority, the Government of Bombay was initially less willing to allow Calcutta to interfere with every detail of its policy, leading to Temple's somewhat unreasonable charge of continuing profligacy. The first months of the famine campaign were overshadowed by a lengthy dispute over the ideal scale of relief works. The local government was keen to open major projects, including the Dhond–Manmad railway through Ahmednagar, but was denied permission to do so on the grounds that they would probably not be finished before the famine had ended, creating an unjustifiable expense.[30] This decision, reflecting the central government's unwillingness to acknowledge the extent of distress, was only overturned in February 1877 when Bombay had the wit to argue that small works were costlier and less efficient because of the higher ratio of administrators to workers and reliance on revenue officers rather than the PWD.[31]

The Bombay Government followed this argument through by keeping its own tight rein on the implementation of relief in Ahmednagar, withholding sanction from small works until they had received a budget estimate. Thus red tape delayed local attempts to meet the demand for relief, while works likely to cost more than average were blocked, regardless of their catchment area. This infuriated Collector Jacomb, who persistently opened works in anticipation of sanction, earning Revenue Commissioner Robertson's censure for his 'grave irregularity and neglect' over estimates.[32] In frustration at the failure of his superiors to acknowledge the seriousness and scale of the crisis, Jacomb insisted 'you are aware that such procedure may cause delay and that no people are now employed in agricultural operations'.[33] In November 1876 he reported that 30,000 labourers were already on relief in the district and that works already sanctioned would 'last till about the end of next month.'[34] Though a programme of 40 potential works was approved in response, he was reminded to submit estimates for each one and report on 'the urgent necessity for relief' before opening them.[35] Jacomb replied that the urgency of demand had already overtaken the speed at which sanction could be granted: 'The numbers of people … who came in search of labour as soon as a work was undertaken increased rapidly and it became necessary to think of other projects in addition.'[36]

This ongoing battle was openly over control of relief decisions, with many of Ahmednagar officers' judgements challenged. A typical response followed Spry's report that crop failures had created demand for a work on the Newasa–Sonai road, costing 3000 rupees.[37] Though cheap enough, this was only sanctioned alongside a stinging rebuke from the PWD: 'Sufficiently strong ground has hardly been assigned for the commencement of this work. No new work should be commenced unless there is a *strong* demand for relief, or *until the people begin to show signs of distress.*'[38] Thus, revenue officers were expected to distinguish an excess supply of labour from need, a pernicious task that legitimised suffering. As this example shows, attempts to make such distinctions were often translated by the hierarchical decision-making process into the selection of cheap works, rather than needed ones. District officers eventually responded by adjusting their requests to increase the chances of their approval. Woodburn supplied Jacomb with a list of small works which he, echoing the Government of Bombay's original argument, did not see as efficient to manage 'which I propose should be carried out, in the event of no large work being sanctioned by Government'.[39] Later, Executive Engineer Howard submitted a revised estimate for the completion of the larger Loni–Bari road work after Bombay had rejected his original one, commenting frankly 'I very much doubt whether this amount will suffice, but I have reduced the Estimate in accordance with your wishes.'[40]

The Government of Bombay also sought to keep its relief expenditure down by charging the cost of any works of sole value to Ahmednagar district to local coffers, though they still required sanction. Chapter 2 showed that the Local Fund system was only sufficient for limited district infrastructural projects anyway. The pressure of devolving costs to it during the famine – based on a problematic trade-off between the relief necessity and utility of projects, which is discussed later – left district funds hopelessly overstretched and unable to effect even basic maintenance, let alone improvement, for many years afterwards. This strategy demonstrated the Bombay Government's awareness of the subsequent importance that would be attached to their famine expenditure. Even before Temple had taken up his post, he lowered the total budget estimate for the Presidency to £1.16 million. This was raised to £1.29 million when the summer monsoon failed again in 1877.[41] When the Bombay Government announced its final spending figure in 1878, it came to £1,591,579. That was not an unreasonable overshoot, but they attempted to show that this figure was not the same as the net cost of the famine, as shown in the breakdown in Table 5.2.

These figures are problematic in several ways. As seen in the previous chapter, land revenue remission was proscribed, with much of the small cost recorded here coming early in the famine and that written off later not included. That it should be dwarfed by extra administrative costs, largely due to the need for British officers to travel and correspond constantly, reflects the meanness of that policy. When juxtaposed with a £350,000 net profit on the import of foodgrains on government-owned railways, the Bombay Government laid itself open to the charge of benefiting from the famine.

While the advantages of irrigation and transport links constructed during the famine may or may not have been shared by the population, it was spurious to discount them from the cost of famine on that basis. Given that works financed by Local Funds tended to have little utility because of their low quality, excluding them entirely from this calculation was even more suspicious. Most striking of all, the figure admitted here for Public Works does not include the major one in Ahmednagar.

Table 5.2 Famine balance sheet, Bombay Presidency

Debit	Pounds	Credit	Pounds
Total Public Works expenditure (not including Local Funds)	1,001,800	Value of works of permanent utility	390,000
Gratuitous relief	120,374	Sale proceeds of government hay/grain	4,313
Preservation of agricultural stock	13,824	Extra profit on guaranteed railways	670,000
Revenue remission	25,581		
Loss of ordinary traffic on guaranteed railways	320,000		
Additional costs of administration	110,000		
Total	1,591,579	Total	1,064,313
Net cost			527,266

Source: General Report on the Administration of the Bombay Presidency (GRABP), 1878, p. xvii.

The Bombay Administration Report explained:

> The cost of the Dhond and Manmad State railway (£87,152) has not been entered in the above calculations, as, although it afforded employment at one time to about 30,000 labourers who would otherwise have been in receipt of relief elsewhere, it was continued as an ordinary work to relieve the pressure on the Great Indian Peninsular railway.[42]

While this can be seen as further acknowledgement that the line was not designed to benefit the district, excluding relief works on the debit side because they were useful, before adding others on the credit side for the same reason, reveals the provincial government's desperation to produce figures that showed their frugal credentials.

This priority also affected the availability and ethos of relief works, with every project challenged and individual applicant tested. Temple was sensitive about comparisons between his Bombay campaign and that in Bihar in 1874, which successfully prevented mortality but cost around £6 million. However, when asked what level of expenditure he would recommend in future, Temple admitted that 'subsequent experience in 1877 shows that ... the life of the people could not be saved for a less sum', at least *per capita*, than his Bihar costs.[43] In 1876–78, he concluded that Bombay's final accounts were 'favourable', showing that 'every endeavour was made by the relief officers to maintain economy'.[44] As for the population, 'while lamenting the loss of those who have perished, and pitying the misery and privation endured by so many, we may hope that the people have learned a hard lesson of self-dependence'.[45] Thus he acknowledged that reducing costs had been at the expense of lives.

This attempt to scale back state obligations did not refer solely to the famine period. Immediately afterwards, a heated debate broke out about local responsibility for relief. Once again, this was prompted by John Strachey, who calculated that the total unanticipated cost of state famine responses had been £16 million between 1866 and 1878.[46] Given that famines could no longer be seen as 'abnormal or exceptional', it was necessary to raise extra revenue on an annual basis to finance their relief.[47] It is interesting to note that he quickly ruled out a grant offered by Britain – in order to maintain 'financial independence' – and also a re-introduction of income tax.[48] While the former struck the *Parsi* newspaper *Jame Jamsed* as folly when India simply could not afford to manage alone,[49] the Government of Bombay consistently argued that only income tax would be sufficient to cover famine costs.

Strachey instead proposed an additional cess to create local famine funds, on the basis that communities should protect themselves. The government's responsibility, as his brother Richard had declared at the Simla meeting, was not to feed people, only to 'see that they mutually aid one another'.[50] A similar suggestion had been made by the Orissa Famine Commissioners in 1867, provoking D'Oyly in Ahmednagar to declare that 'Local taxation has increased greatly of late years and is very unpopular.'[51] When the question was again circulated to officers in Bombay by the 1880 Famine Commissioners, Robertson launched into a fierce attack of what he saw as an extension of the already 'ruinous' policy of bankrupting existing Local Funds before spending provincial monies on relief, particularly in districts which had long been ignored: 'Till the Local Fund system was established, government did next to nothing for the country. No roads were made or repaired, no schools or dharmsalas built, no wells or tanks repaired; in fact neglect of the most glaring description was the order of the day.'[52] Local responsibility for both development and relief was widely seen as an excuse for such neglect, rather than a policy that could be defended in practice or principle.

Typically, Strachey paid little attention to local condemnation. Though he did acknowledge that 'The justice of imposing new burdens on the agricultural classes will possibly be ... questioned. ... There may be some parts of India where the land revenue is so high that it would be unwise to make fresh demands', he remained opposed to any form of redistribution of wealth, even in crises: 'The mere fact that the agricultural classes constitute by far the greater proportion of the population, and, when Famine occurs, form the great majority of those who require relief, is alone sufficient to show that these classes ought to pay their quota of the sum required for their own protection.'[53] This argument rested, significantly, on the assumption that revenue-paying *ryots*, as much as the rural landless, depended on relief, which was contradicted by Temple's conclusions from the Bombay campaign. Thus peasants were to be charged for relief they could not afford and had been denied. Fiscal responsibility was to be devolved as low as possible while key decisions were made at such a high level that they ignored the particular circumstances of a whole presidency, never mind a district. This might have been expected to raise hackles in Bombay had it concerned anything but the chance to increase local revenue. In April 1880, however, to even the Government of India's surprise, a Bombay Government Resolution called for the imposition of a 'special famine rate' of half an *anna* in every rupee of land assessment, to be spent on 'preventive' irrigation works, rather than relief.[54] This was all the more surprising following a sharp exchange on the subject two years earlier, in which Strachey's condemnation of irrigation as wasteful left Temple furiously – and revealingly – defending himself against the perceived charge of profligacy.[55]

The nature of relief

Temple was thorough in his approach to famine mitigation, insisting – as he had in Bihar – on regular Village Inspection to ensure that the 'helpless' did not die in their homes.[56] In Ahmednagar, however, Jacomb's attempt to ensure this by appointing eleven circle inspectors was blocked on the grounds that over-worked village and *taluka* officers should be able to find time for it.[57] Moreover, the term 'helpless' was meant literally, to exclude those whose 'refusal' of relief put them beyond the state's responsibility, and Temple's most serious warnings about the need for inspection only followed reports of increasing emaciation after the second rain failure in 1877.[58] By then, he had also adopted the common term 'gratuitous' to refer to relief – given without work being required in return at residential feeding stations – which he had called 'charitable' in 1874. Not only did such relief account for just 7.6 per cent of Bombay's total famine budget, as Table 5.2 shows, its very premise was under challenge by August 1877, when the Bombay Government suggested its recipients should also work. Jacomb readily confirmed that everyone seeking relief in Ahmednagar was sent to work unless 'physically incapacitated to do so' and that 'the grant of charitable relief is confined to the worst cases. Moreover I can assure Government that the severest economy has been observed throughout as regards this source of relief in this District.'[59]

Such promises of stringency were in marked contrast to his earlier warning to Ahmednagar *patels* to be aware of their duties in line with Cooper's still extant memorandum:

> Should any person die of starvation in your village you will be held personally responsible. You are constantly to go round your village and when you find any person suffering from extreme want you are at once to give that person sufficient food for one meal and should he or she be able to travel, send him or her off to the Mamledar's station. Should that person from old age, sickness, weakness or any other cause be unable to travel, you will report the circumstance immediately by special messenger to the Mamledar who will arrange for the feeding of the person in your village.[60]

The Government of Bombay ordered that such treatment should not be afforded to travellers, unless they could demonstrate their inability to find food by other means.[61]

Relief works, like all state infrastructural projects, were divided into two categories: those run by the PWD, and those to be carried out by Civil Agency, meaning district revenue officers. The Famine Commission Report recorded total numbers on PWD works in the presidency of 2,463,235,[62] with a daily average of 285,000.[63] The total on Civil Agency works was just 555,469,[64] suggesting they played a relatively small part in relief, though it is likely that many on disorganised local works were not counted. In Ahmednagar, the highest number recorded on relief – in June 1877 – was 59,129.[65] Both types of relief were informed by the colonial priorities of economy and efficiency rather than efficacy, though these were played out in different ways. It was to the population of Ahmednagar's advantage to have one of the biggest PWD works in the presidency, the Dhond–Manmad line, close at hand. Some parts of the district had easier access to it than others, however, and remoter *talukas* suffered after its opening in March 1877 when all other PWD works were ordered to close.[66] Robertson had argued for this but Ashburner, his counterpart in the Northern Division, protested at the folly of turning over 8000 men off the Sinnar–Goti road in the north of Akola *taluka* to travel over 120 miles to the nearest railway work in 'the unknown country of the Deccan, where they have heard there is real famine'.[67] While this shows an unusual sympathy with people's views, Ashburner was careful also to argue from a practical point of view, making administrative priorities explicit in the process:

> I anticipate great difficulty in drafting the people to the Dhond line; besides their well known repugnance to leave the neighbourhood of their homes, there is the real difficulty of feeding them on the march and the necessity of the cultivators at least returning to their homes before the monsoon, so that we shall lose about 20 days of their labour, during which time we shall have to support them and shall therefore not get more than a month's work out of them on the Chord line, before the monsoon.[68]

Any benefit of the railway for the local economy was offset by the use of exceptionally cheap labour at a time of crisis. The Bombay Government sought to ensure that PWD works gave value for money. While some, such as Temple, saw relief works as a win–win solution, with every irrigation tank or mile of track reducing the risk of future food shortages, it has been seen in Chapter 2 that the Dhond–Manmad line was built for quite different reasons, with the state the prime beneficiary of its construction. This was not untypical. Sanjay Sharma has shown that famine labour was used in 1861 to enhance the power of the state itself by constructing barracks and police stations.[69]

PWD relief works amounted to a temporary employment guarantee scheme, demanding a full day's hard labour in the heat of a drought for the prescribed wage. Thus the 'famine of work' was responded to but there was little relief for labourers from the wider physical effects of the famine. PWD officers – already having to make do with an unskilled labour force – were reluctant to employ anybody without the strength to complete their standard daily tasks. Ahmednagar newspaper *Nyaya Sindhu* reported that 'subordinates entrusted with the management of the works, refuse to employ any but the strongest men as labourers'.[70] Though the PWD told its men that this was 'manifestly inconsistent with the object for which the works have been opened',[71] famine workers were still expected to do the jobs of experienced PWD labourers on much lower wages. This conflicted with the requirement to demonstrate their need, making them vulnerable to both administrative confusion and physical abuse. The *Poona Sarvajanik Sabha* claimed that 'Blows and boxes are given on the chests of the labourers, and if anybody complains of the pain caused by the blows, he is rejected as infirm.'[72] Those judged too weak by the PWD also had to walk a long way to Civil Agency works, as Jacomb insisted they should not be close to the railway.[73] The *Times of India* condemned the muddle and reported deaths as a result of 'weary waiting' to be employed.[74] Robertson suggested that it should be local revenue officers who decided where workers should be sent, and admitted that tension between departments amid the chaos of the famine had caused communication to break down: 'When the arrangements are good and officers are working harmoniously such disputes should never arise.'[75] The Bombay Famine Code, published eight years later, felt it necessary to remind PWD officers that 'whereas ordinarily the primary object is to get the work done as cheaply and effectually as possible, with no special regard for the well-being of the labourer, the principal object in famine is to keep him alive'.[76]

This was made harder in the 1876–78 famine by the unsanitary and unhealthy conditions on relief works. It is interesting to note the Bombay Government's language when they demanded 'the *concentration* of as much relief labour as possible on large works, such as the Dhond and Manmad Railway'.[77] Though still two decades before the Boer War made the term notorious, the notion of famine works as concentration camps reflected the common adoption of military rhetoric in the conduct of the famine campaign. Most strikingly, Elliott justified increased mortality rates in 1876–78 to the Simla meeting by declaring, 'The campaign against famine, like other wars, if properly conducted must have its butcher's bill.'[78] Famine workers were treated with as much hostility as prisoners of war and suffered similar epidemiological consequences. Excess numbers were a common problem, on both Civil Agency and PWD works, as a result of the government's reluctance to sanction new works. When the Bombay Sanitary Commissioner, T. G. Hewlett, attempted to restrict groups of workers to 500, kept at least a quarter of a mile apart 'on account of the great danger of overcrowding, and of inflicting injury on persons residing in one encampment, by exposing them to the filth created in the next',[79] his orders were contested by Colonel Merriman, the more influential PWD Chief Engineer.[80] The two also disagreed over the necessity of medical supervision of grain distribution, with 'so much bad grain now being sold' and the provision of clothing to labourers, without which 'the sick list will be very heavy'.[81] Even when Hewlett proposed camp hospitals, Merriman held that they should be erected only when required.[82] Major Kennedy, Secretary of the PWD in the Government of Bombay, invited the pair to sit down together and agree simple guidelines, but made his preference clear in pointing out that 'It is not so much for consideration what *ought* to be done if time and means were more available, as what *can* be done under the present emergency and great pressure.'[83]

Hewlett also spelled out the ideal procedure in the event of the most serious threat on relief works: 'Should epidemic disease, such as cholera, break out, the patients should be treated *not* in the general hospital, but in a temporary hospital' with camps to be abandoned completely should more than three cases of cholera occur, and all workers marched 'at right angles to the wind'.[84] Of course, the latter would not have helped and was impractical on a railway line, while the former was academic after Merriman's refusal to create hospitals. Nonetheless, Hewlett's anticipation of cholera in such conditions made good sense. There were periodic reports of minor outbreaks on Ahmednagar camps and the disease was widespread in the district throughout the famine.[85] From August 1876 to May 1877, 3556 cases were recorded, with 1585 reported deaths.[86] In the following year, 1954 of 4265 cholera patients died, and Stewart reported a renewed outbreak in June 1878.[87] This compared relatively favourably with a death toll of 41,250 from cholera in the presidency from January to September 1877 alone.[88] Though the epidemic had started before the famine, rather than because of it, relief conditions considerably exacerbated it.

The Government of Bombay was well aware of the health dangers inherent in concentrating relief on a few large works but they failed to pay sufficiently serious attention to disease in the famine campaign. In Klein's words, 'The policy of averting death by starvation was not entirely unsuccessful, but relief was insufficient to ward off famine-induced diseases, which killed millions.'[89] Along with Arnold, Dyson, Lardinois and Whitcombe, he highlights the particular significance of cholera as a contributor to mortality during the 1876–78 famine, because workers had travelled from far and wide, drinking from nearly dry wells and surface water, to crowded camps where sanitary considerations had taken second place.[90] Even on smaller Civil Agency works, Hewlett's advice had not been passed on, judging by Spry's reaction when cholera broke out on a tank excavation work in Newasa. He ordered those affected to be transferred immediately to other tanks, preferably far away.[91] While cholera's association with water had not yet been accepted in India, it was certainly known that it was spread by the movement of its victims, hence Hewlett's recommendation of their isolation.

Tests of eligibility for relief

From the beginning of the famine, the cost-conscious Bombay Government was determined that only the desperate should be offered work, claiming the desire to prevent the population from being 'demoralised' by easy relief. The four specific tests applied – appearance, labour, distance and wage – ensured relief would only be taken by people showing visible signs of hunger who were willing to work hard and long for a pitiful wage, far from their homes. They relied on self-selection, with the exception of the appearance test. By excluding those with obvious assets, such as jewellery, decent clothing or even, in Temple's words, 'splendid physical condition',[92] this was arguably the most invidious test of all, requiring people to surrender their security before being helped. As the newspaper *Rast Goftar* noted, forced asset sales were far more likely to create dependency than would easier access to relief.[93] Women were particular disadvantaged in Ahmednagar by Jacomb's insistence on automatic disqualification for 'possession of valuable ornaments'[94] although his assistant, Hamilton, argued that 'no labourers in this district ever have more than copper imitation jewellery'.[95] Given the Government of Bombay's desire for control and consistency, it is surprising that so much was allowed to depend on local officers' impressionism, and the appearance test was dropped from the subsequent Famine Codes.

Though the other three had far greater longevity, all four were tests in more senses than one. As the Government of Bombay, with all India's eyes upon it, sought to devise famine policy as it went along, many strategies were unashamedly experimental. This was especially true of the distance test, the idea for which was credited to Jacomb in the Famine Commission Report.[96] The suggestion that providing relief in people's own villages was too generous had in fact first been made to him by Spry,[97] and the capacity to reduce numbers of relief applicants by requiring them to travel accidentally discovered by Fforde. When he reported that the transfer of 1000 labourers from Parner *taluka* to the northern Kopargaon–Newasa road had resulted in 375 going home instead,[98] Robertson took this as evidence that they were 'not at present in want of work' and suggested that 'wherever without hardship the rule, of transferring men to some distance, can be enforced, it should be strictly adhered to'.[99]

The government was happy to agree, ruling that no one living within 10 miles should be allowed onto any work. They recognised that the need to travel was 'repugnant' to labourers, but argued that 'distasteful' working conditions were an advantage, successfully proving that those who accepted them were in need.[100] By contrast, one of the *Nizam*'s senior famine officers, Maulavi Mushtaq Husen, opposed the distance test, so as to preserve people's non-travelling habits.[101] This objection was echoed by Major-General Wilkins of the PWD, who offered a revealing insight into British fears that too strict terms may affect their legitimacy: 'Famine-stricken people should never be compelled to go anywhere. They are sure to suffer in some form more or less, and they will attribute their sufferings to their rulers if they are forced against their will to follow any particular course.'[102] This was not the only concern. Children, the old and, to a lesser extent, women were more likely to be left behind because of the requirement to travel, unprotected and in need of charitable relief. Empty homes also attracted looters. When labourers did travel in large numbers, they increased the spread of disease. Hamilton also reported that the distance test was 'troublesome to carry out' and unnecessary on top of the wage and labour tests, which suggests that it was not popular with all administrators in Ahmednagar.[103] Above all, sending *ryots* away from their fields meant that they would not be able to return to cultivation immediately when rains returned. Thus, again, the test ran the risk of increasing the dependence on relief it claimed to eliminate, and even of prolonging the famine by inhibiting local production. This danger was recognised by Temple, who argued that it should not be applied in the monsoon period and with discretion at other times,[104] but Ashburner went further after joining the Government of Bombay: 'The distance test should be abandoned. It works very cruelly and retards the recovery from famine.'[105]

The labour test, on the other hand, was far more consistently applied. Based on Benthamite principles, putting the destitute to hard labour was effectively a temporary application of the English Poor Laws to India,[106] and seemed obvious to most British officers. Even today, many relief policy-makers argue that cash (or food) for work is more appropriate than food handouts. Confusion reigned, however, over the nature and amount of work required on the PWD and lighter Civil Agency works. As late as February 1877, Spry complained that he had to leave decisions to the discretion of his *mamlatdars* as 'I have not yet received any standard of the amount of task work to be exacted from different grades of labourers.'[107] The government had attempted to explain that 'a full day's work, according to their sex and age, shall be exacted' but that tasks should be at least 25 per cent below normal on PWD works, and 50 per cent below under Civil Agency.[108] After further uncertainty, this turned out not to be a proportion of the daily PWD task in non-famine times but of workers' 'individual capacity'.[109]

As a result, the policy was applied inconsistently, reflecting the ambivalence of local officers. Even Howard, the PWD's man in Ahmednagar, initially stressed that 'as the works are for *relief* hard tasks are *not* to be set',[110] though he also ordered that 'no hesitation need be felt in cutting half a day's pay from any *able-bodied people* who through idleness or mischief will not work'.[111] Again, it was difficult to determine what would be hard for workers unfamiliar with such tasks. Superintending Engineer Jenkin Jones argued, ironically, that while railway work like embanking or cutting required special skills, road works were ideal for unskilled labour. His description of the main task involved, breaking stones, did not augur well for the agriculturalists: 'The new hands knock themselves about a little at first as might be expected, but they soon get into the way of hitting the stone instead of their own bodies.'[112] Small wonder that Hewlett criticised the entire concept of a test involving 'inappropriate' labour, especially for artisans and women.[113]

While most colonial famine records are gender blind, rarely identifying differences between the experience of male and female workers and using the male pronoun for both, there were many women present on famine works. Where the distance test was applied, families often travelled together to take relief, and when they split up temporarily to ensure an income while still tending their holdings, women were more likely to end up on the demeaning works, just as Stephen Devereux shows they are today.[114] Indeed the Famine Commission Report recorded an 'unusual proportion' of women and a 'large fraction' of children who were 'unfit for labour' in camps.[115] References to these as 'dependants' were pejorative and those over the age of seven were expected to fulfil tasks in proportion to their wage. This made women with children on PWD works particularly liable to see their wages docked for failure to meet the hard tasks set. Candy described how emaciated labourers at Dhond 'had, every time they carried a basket of earth, to climb out of a pit six feet deep, and then mount a steep bank 20 feet high. This of itself I call a task likely to drain the nervous force of the Coolies during the intense heat of the day.'[116] Candy made no mention of their gender, but according to Jocelyn Kynch carrying was the archetypal female task.[117]

What Candy did argue was that failure to complete set tasks in such circumstances demonstrated not idleness but excessive demands. Yet fines for short work meant many labourers received only the equivalent of the reduced Civil Agency wage, on which their condition was deteriorating.[118] The PWD indignantly insisted that reduced pay meant that tasks had not been performed,[119] but Candy's equally heated response makes it clear that what had been intended as punishments for refusal to work had become a device for short-changing hard workers to a dangerous degree: 'What I contend for is that Civil Agency rates cannot purchase enough food to restore the exhausted energies of coolies who have been doing such work as I have described was being done at Dhond and Ghospuri, even though they may have not succeeded in performing the full task demanded.'[120]

While this case of excessive task-setting and wage-cutting may not have been the norm, it reflected the general desire of the PWD to get as much work, rather than relief, done for their money. This was not even efficient, as low food intake reduced workers' productive efficiency.[121] Moreover, the severe labour test policy was so discouraging that at one point Candy reported 'the engineers are crying out for labourers'[122] and the government were obliged to pay skilled workers at full rates when they refused to accept famine wages set for 'common coolee work'.[123] Nonetheless, Sub-Engineer Knight declared 'Famine Relief Work can be carried out profitably with good management.'[124] This was backed up by the belief of Poona's Executive Engineer, Captain Seton, that works completed by relief labourers should be cheaper than those at other times 'in direct proportion' to the lower wages, implying the expectation of a full task from all famine labourers, including women and children.[125]

By contrast, the *Nizam*'s revenue minister, Sir Salar Jung, justified road work costs over four times higher during the famine on the grounds that weak inexperienced labourers could do so much less.[126] This was borne out in Ahmednagar when Hamilton successfully requested sanction to increase the budget on a Sangamner earthwork because 'The labourers on this work being mostly old men and women with a vast number of small children the work is costing 12 annas per 100 cubic feet instead of 8 annas which was my original estimate.'[127] As the wages were fixed for each category, this suggests that Hamilton, unlike Seton, did not expect tasks to be completed on the same scale as wages. Earlier, he had argued similarly that 'if work is exacted according to physical capacity all labourers should receive alike, whether under the PWD or under the Collector'.[128] This amounted to a radical critique of Temple's key policy of drastically differentiating Civil Agency and PWD wages. The Government of Bombay, however, maintained that part of the aim of task work was 'to obtain a return for the money the State is compelled to expend on relief'.[129] Testing people's need to such an extent by this means came too close to exploitation.

The most significant and controversial test in the 1876–78 famine campaign was that of low wages, especially after Temple's intervention.

He created a categorical distinction between PWD and Civil Agency works, ordering that the wage on the latter should be the cash equivalent of 1 lb of grain per day for a working man. This provoked furious protests from many quarters, most notably on medical grounds by the Sanitary Commissioner of Madras, Robert Cornish, who argued – in similar vein to Candy – that this was insufficient to sustain the hungry's labour.[130] This prompted the Government of Madras to reject the 1 lb wage after six weeks, which Temple argued was not a 'proper trial'. Significantly, he added that he had never claimed that it was 'definitely sufficient', only that it should be tried and retained where it worked.[131] The wage was thus openly experimental. Cornish's Bombay counterpart, Hewlett, however, refused to oppose Temple, apparently for political reasons.[132] This altercation and its relief consequences – particularly in Madras – have been well covered in literature.[133] The aim here is to examine the application of the wage test, and particularly the 'Temple Wage', in Ahmednagar.

In the early stages of the famine, the Government of Bombay had set a fixed cash wage of two *annas* per day for men, one and a half for women and one for children capable of working.[134] At this point, *jowar* was still around 16 lbs for a rupee, so a man could buy 2 lbs of grain with his daily wage. In December 1876, under Temple's influence, they instead adopted a sliding scale, so that the wage depended on the rapidly changing price of food. Men on PWD works were to get one *anna* plus the value of a pound of 'a medium quality of the cheapest grain in ordinary use among the labouring classes of the District',[135] women six *pice* (half an *anna*) plus a pound and working children either six *pice* and half a pound, or a pound.[136] It is notable that, at this stage, concern over the ability of local dealers to supply the works allowed for the possibility of partial payments in grain. The new sliding scale is given in Table 5.3.

As grain reached its high famine levels of 8–9 lbs a rupee, therefore, adult male labourers could afford only $1\frac{1}{2}$ lbs. This was at least consistent, however, with prevailing market rates for unskilled labour in Ahmednagar city, which had dropped in the famine years, as shown in Table 5.4. Concern that this contradicted the stated principle that 'relief should be granted to none but those … unable to support themselves in any other way',[137] prompted Temple to reduce the wage further on Civil Agency works in January 1877, so that relief wages would 'be to the extent only of a bare subsistence'.[138] On those works run by collectors and their assistants, the only supplement to the equivalent of a pound of grain was six *pice* for men and three *pice* for women. Working children, and also 'disobedient' labourers who failed to work, would receive

Table 5.3 PWD relief wages payable (*annas* and *pice* per day), Bombay Presidency, December 1876

Price of grain (lbs per rupee)	Value of 1 lb grain	Total pay		Child's pay if 6 *pice* +½ lb	Child's pay if 1 lb
		Men	Women		
8	2-0	3-0	2-6	1-6	2-0
9	1-9	2-9	2-3	1-5	1-9
10	1-7	2-7	2-1	1-4	1-7
11	1-6	2-6	2-0	1-3	1-6
12	1-4	2-4	1-10	1-2	1-4
13	1-3	2-3	1-9	1-2	1-3
14	1-2	2-2	1-8	1-1	1-2
15	1-1	2-1	1-7	1-1	1-1
16–18	1-0	2-0	1-6	1-0	1-0
19	0-11	1-11	1-5	1-0	0-11
20	0-10	1-10	1-4	0-11	0-10
21–22	0-9	1-9	1-3	0-11	0-9
23–26	0-8	1-8	1-2	0-10	0-8
27–30	0-7	1-7	1-1	0-10	0-7
31–32	0-6	1-6	1-0	0-9	0-6

Source: Attachment to Government of Bombay Resolution No. 268 C.W.-1038, 13 December 1876, Famine Vol. 62, p. 121.

Table 5.4 Average market rates for unskilled labour (*annas* per day), Ahmednagar city, 1869–78

Year	Average daily wage
1869–70	2–3½
1870–71	2–3
1871–72	½–3
1872–73	4
1873–74	4
1874–75	4
1875–76	'up to 4'
1876–77	'up to 3'
1877–78	2½
1878–79	2½
1879–80	3
1880–81	3
1881–82	3

Sources: Boswell to Havelock, No. 1645, 20 July 1874, p. 117; Boswell to Havelock, No. 2132, 20 July 1875, p. 449; Stewart to Robertson, No. 1046, 22–4 July 1878, p. 285; King to Robertson, No. 4161, 20 July 1880, p. 18; Elphinston to Robertson, No. 5730, 20 July 1882, p. 63.

the equivalent of three-quarters of a pound of grain, or half a pound and three *pice*.[139]

Even those opposed to Temple's wage accepted the principle of offering only subsistence rates, but it was questionable whether that could be maintained by a wage deliberately set below market rates during a 'famine of work'. As the state was by some way the biggest employer in the district during the famine anyway, it was recognised that it was barely necessary to set wages below market levels already pulled down by its own participation.[140] Indeed Himmelfarb has suggested that public relief works with higher wages are as likely to have the knock-on benefit of raising levels of pay elsewhere as to attract more workers.[141] Depressing though the relief wages were, it can be seen from Table 5.4 that labour rates took a long time to recover after the works had closed. This raises the question of why private industry did not take advantage of the supply of cheap labour during and after the famine in Ahmednagar, with the answer lying in the general lack of investment, recession and credit squeeze seen in previous chapters. Given such a chronic lack of alternative employment in the district, the reduction of wages was not to protect private employers so much as to reduce the burden on the state.

A pound of grain is not strikingly little, however (depending on its quality), and Temple's main defence amid the furore was that it was not the wage itself but short payments through shoddy administration or fraud that were to blame for any deterioration in workers' condition. While it was typical to pass the buck to predominantly native administrators in this way, there was indeed an array of problems in managing relief payments. Jacomb listed obstacles to survival on relief thus:

> Notwithstanding all the precautionary measures that may be taken, the probable disadvantages that the labourers suffer in being cheated in the payment of their wages according to the prevailing rates, the difficulty they experience in procuring grain at the rates at which such wages are calculated, the loss of deduction on account of the inability to render full tasks, the charges for credit, the results of accidents and sickness &cc, the exhaustion of private resources (no well to do people now being on the works) and want of clothing and last not least the uncertainties of prices and their determination.[142]

Given the total numbers on relief and both spatial and temporal fluctuations in prices, the sliding scale system was too complicated and failed to allow properly for workers' needs other than grain – such as salt, chillies, cooking materials and fuel. The arbitrary differences between male and female wages also failed to take account of specific needs, such as pregnancy. Ahead of his time, Robertson noticed the particular danger of paying so little to lactating women, for whom 'the food is not sufficient to enable them to provide proper nourishment for their babies'.[143] Indeed no allowance was given to parents of children too young to work until January 1877 and this was fixed at three *pice*, not sliding. When prices went up, it was inevitably reported that this could not sustain a sufficient diet.[144] Similarly, Temple specifically ordered that wages should not be paid on Sundays, when no work was done, requiring the 'daily' allowance to be stretched further – although holidays with pay were allowed for New Year and the Queen's birthday.[145] Taking this into account actually made the Civil Agency wage in Bombay lower per week than Madras, where it was so hotly contested.[146] Sunday wages were finally permitted in Bombay only after a second successive monsoon failure had been established, in August 1877.[147]

For practical reasons, wages were rarely paid daily anyway, but weekly or fortnightly, leaving workers to husband their resources in the face of potential price rises. To make things worse, when sliding scale calculations left fractions of *pice* in the daily wage, they were only rounded up after being aggregated for the whole period.[148] A larger difficulty for workers came when they were given less than the prescribed daily wage, and it was by no means only native overseers who were guilty of this. While touring the railway work with Temple, Robertson reported being 'met on all sides by complaints as to the small amount of wage received during the past week, several gangs advertising they had only received 4 annas per man'. On inspecting pay sheets, he discovered that this was indeed what had been given to cover six days, albeit including two on which labourers refused to work after being fined by the line's Superintending Engineer, Izat, for non-completion of tasks. Robertson 'pointed out to Mr Izat that a payment of one anna per day was totally insufficient to sustain life' and requested him not to impose fines that took wages below the Civil Agency rate.[149] This was a case he had seen. Countless over-strict charges for short work are likely to have been levied throughout the district. Newspapers occasionally made accusations of deliberate short waging by relief overseers, but few concrete cases emerged in Ahmednagar. As far as British officers were concerned, the problem was more likely to come from *banias*, who sold bad food and took advantage of workers' lack of knowledge of prices, as seen in Chapter 2. Had those revenue and PWD officials willing to do so in Ahmednagar been allowed to supply grain to camps for the sake of administrative convenience, these problems would have been avoided.

The re-categorisation of works – and wages – according to the assumed levels of work being done also created considerable confusion. Assistant collectors complained that 'the system of double agency + double rate cannot yield good results unless there be a wide distinction in the class of labourers employed under each agency'.[150] It was pernicious that payment became dependent on who managed a work, rather than tasks completed. When Jacomb concurred, however, that it was unfair that workers on the Sakur–Rahata road work should receive lower wages 'if as it appears, they are doing the same rate of work as is done on the Public Works Department works',[151] he was scolded by the government: 'Proposals to pay persons employed under Civil Agency at Public Works rates must not be repeated.'[152]

Self-evidently, the rationale for the 'Temple wage' was to minimise the costs of the relief campaign. The chief effectiveness of the new wage was as a test, with thousands choosing not to accept work in those conditions at that rate – including many already on relief. In Temple's final minute he expressed open pleasure that numbers on works had peaked at 10 per cent of the Presidency's population, incredibly claiming that this reflected good relief management as well as the well-being of the population 'after many years of careful revenue settlements and just administration'.[153] Rejecting this 'showy special pleading', the British-run *Bombay Gazette* challenged the Famine Commission to investigate the reasons for the 'extraordinary reluctance' of so many people to take relief.[154] By requiring poor peasants either to accept a sub-subsistence income in return for labour or fend for themselves, the Temple wage was as much to blame for exacerbating the famine as any British policy.

Local criticism was loud and long, though less effective than in Madras. The *Poona Sarvajanik Sabha* warned 'The new sliding scale of wages threatens to inflict death upon many hundreds of persons by the slow process of gradual starvation'[155] while *Dnyan Prakash* condemned persistence with experimental 'theory against the strongest arguments of science and experience'.[156] The Government of Bombay ignored such native views, insisting 'Experience has shown that these rates ... were enough to keep the labourers in health; whilst insufficient to attract persons not entitled to claim relief.'[157] District officers' opinions were ostensibly taken more seriously. The Famine Commission reported that no problems had been found with the PWD rates, and 'an almost general consent that the Civil Agency wage was also sufficient'.[158] Jacomb, however, expressed concern at the new rates from the start, asking his assistants to 'carefully watch the effects of their operation', adding 'should you detect signs of starvation which you are satisfied is the result of such orders, I request you will lose no time in reporting the matter in full to me'.[159] After two months, he concluded that 'the Civil Agency wage is only sufficient for bare support without any margin whatsoever for accidents'[160] and that 'a slight increase if possible might with advantage be allowed at once instead of hereafter when such a step will probably be an absolute necessity'.[161]

Assistant collectors' views varied. Spry was unconvinced, but Hamilton was certain that the wage had 'proved to be insufficient for the support of those labourers who are compelled to work at a distance from their homes or who have no other means of subsistence'.[162] He even conducted an experiment to prove it: 'I procured yesterday an anna's worth of bajri flour and had it made into the ordinary native "bhakari." Four cakes resulted which I am positive would hardly keep a working man alive.'[163] Candy, meanwhile, challenged Temple's strategy and philosophy more directly:

> What policy have the Government fixed upon? Do they wish to keep the numbers on the Famine works at a minimum and leave the rest of the people to shift for themselves – or will they reward a long suffering law abiding people by granting their request in the matter of wages[?][164]

More senior officers also criticised the wage. The previous governor, Wodehouse, suggested that his successor's justification was based on 'more or less cursory' relief inspections, concluding 'I feel that we now run much risk of acting with undue severity.'[165] When the Government of Bombay met, after Temple's departure, to plan its Famine Code, Ashburner declared 'It is bad economy to give *labourers* a bare subsistence ... I believe they would have worked better and more economically on a higher scale of food.'[166] The Bombay Famine Code recommended the significantly higher wages in Table 5.5.

It is curious that the Famine Commission Report recorded a consensus that the wage was adequate in Bombay Presidency, given such criticisms. Replies to its questionnaire do show that local officers in Bombay were less publicly critical of the Civil Agency wage – and of Temple's policies generally – than their counterparts in Madras.[167] This was surely due to his presence as Governor. As it was, Assistant Collectors like Hamilton – and indeed Collectors Jacomb and Boswell – ran the risk of being bypassed for promotion by less critical colleagues like Spry, Woodburn and Stewart. Individual expressions of dissent were also isolated by such summaries of consensus or read selectively. For example,

Table 5.5 Famine relief wages and rations (pounds of *jowar* or *bajri* for which the cash equivalent should be given), Bombay Famine Code (1885)

	Maximum	Minimum
Men	$2\frac{1}{4}$	$1\frac{3}{4}$
Women	2	$1\frac{1}{2}$
Children	$1\frac{3}{4}/1\frac{1}{4}$	$1\frac{1}{4}/1$

Source: Bombay Famine Code (1885), p. 29.

Hamilton told the Commission: 'I am quite opposed to the one pound theory. I do not think any actual deaths from starvation occurred amongst the labourers under me to whom only one pound of grain was given, but they had no strength to work properly, and the birth-rate was abnormally low.'[168] This was taken as confirmation that the wage was sufficient to preserve life. The reference to low birth-rates is significant in the light of Tim Dyson's argument that fertility rates are a more reliable measure of famine than mortality.[169]

It was not unknown for individuals to alter their perspectives when their position changed within the colonial edifice. The best example of this was Temple, whose only consistency between the Bombay campaign and that in Bihar in 1874 – where he remitted large amounts of land revenue and provided $1\frac{1}{2}$ lbs of grain per man close to homes, purchased and transported by the state; thus containing disease, maintaining cultivation and shortening the famine – was in his willingness to follow respective viceroys' orders.[170] Once the Bombay famine was over, and a cheap campaign achieved, even he admitted that 'it may be prudent for a time to allow a somewhat larger ration ... upon the plain ground that if perchance poor people do not get all they are entitled to, still they will receive enough to sustain life'.[171]

More locally, Robertson appeared to develop new priorities upon his appointment as Revenue Commissioner in November 1876. Days earlier, as Collector of Dharwar, he had responded furiously to the new PWD wage scale:

> So convinced am I of the very grave error in the rate fixed by Government that I have ventured to take the step, till further orders are received, of directing different rates of wages to be paid ... to prevent a national calamity, vizt, the country being filled with starving men, women and children.[172]

Robertson remained unhappy with low wages, but his handling of the issue changed markedly on his promotion. Accompanying Temple to inspect the railway at Dhond, he found 'a rather large number of very distressed people' whom he quickly separated from the PWD workers for Temple's benefit, allowing the governor to blame *patels* for their inadequate village inspection. On enquiry after Temple had gone, it transpired that 'all these persons have for some time past been either working on Civil Agency works or hangers-on dependent on those working on Civil Agency works'.[173] In his report, Robertson condemned the lower rate,[174] but his unwillingness to let his governor see or hear this helps to explain why Temple remained convinced that his wage was harmless till after the famine.

Robertson seems to have felt, like many British officers, that respecting superiors was more important than communication of unwanted information, and still less opinion. Indeed, when Candy later reported similar problems on the same works, Robertson not only criticised him for doing so, but publicly disagreed: 'Mr Candy's intentions were good, but he too hurriedly jumped at the conclusion that what he had seen was a proof of insufficient wage and overwork.'[175] Moreover, as Revenue Commissioner, the idea of a collector brazenly contradicting government orders as he had done became anathema, as he showed in condemning Jacomb's attempts to open local relief works without prior sanction. It can therefore be argued that individuals' agency in influencing British famine policy was largely subsumed to hierarchy in such a way that the man at the top – John Strachey – was able to set the agenda according to his own individual perceptions and priorities to a disconcerting degree.[176] What this affected, though, was the formal written policy. While that was the legacy left for subsequent relief campaigns, examination of practical decisions made at the local level will offer better understanding of successes and failures during this famine in Ahmednagar.

Responses to famine policy by *ryots* and individual officers

Having ignored native criticisms at all levels during the famine, Temple took a more optimistic view of *ryots'* agency once the famine was over, writing 'The people never seemed to expect that the State would or could do as much for them as has actually been done. They were unwilling, if they could possibly help it, to come upon relief, preferring to run an excessive risk in searching after sustenance for themselves.'[177] While he attributed this to the 'independent, self-reliant and enterprising' state of the peasantry under the *ryotwari* system – what he called 'moral prosperity'[178] – such reluctance had more to do with the inadequacy of the relief offered. The 'excessive risk' referred specifically to the voluntary discharge of 136,000 workers from relief camps throughout the Presidency in the two months after the introduction of the reduced wage,[179] culminating in a public protest at Mungalwar in Sholapur district.[180] In Ahmednagar, officers reported mass departures from all works as soon as the lower wage was announced, with Jacomb reporting in March that 'the Civil Agency works in the worst talukas have been deserted since the introduction of the new rates'.[181]

Temple was convinced that the mass departures were a concerted attempt to force his hand through 'passive resistance', probably organised by the *Poona Sarvajanik Sabha* and village headmen, whom he claimed had provided 'strikers' with the means to survive during their absence.[182] This only strengthened his determination to maintain the lower wage. Such organisation was never proved and William Digby – albeit a supporter of the *Sabha* – concluded that it was unlikely.[183] The poor state of communications during the famine and the spontaneity and huge scale of departures support this. In Ahmednagar, Spry told leaving workers 'that they are perfectly at liberty to take work or not as they like, and that the Relief works are opened solely for the purpose of giving them a subsistence and not for the benefit of Government, as some of them appear to imagine'.[184] He predicted most would come back soon, 'when they find they can get no good by striking and when necessity presses them'.[185] About a quarter did return, but the fact that 102,000 stayed away indefinitely suggests that they no longer regarded relief works as a viable form of subsistence. According to Candy, some refused it even when they had no alternatives: 'The people say that if they stay on the works drawing the present rate of pay they will die and so they prefer to go to their homes so that if they die they may be among those who know them.'[186] Ignoring this, the Government of Bombay washed its hands of those 'holding out through obstinacy under belief that they will not die, a belief not shared by this Government'.[187] Fforde expressed concern at this notion of ' "dying from obstinacy," which may also mean a sense of injustice'[188] but the government declared that no relief should be granted to anyone 'wilfully and deliberately' refusing to work, thus helping to fulfil their warning of starvation. Candy pointed out that this contradicted the resolution passed a week earlier that names of all those leaving relief should be sent to village officers, who would remain responsible for preventing mortality at home – though in his view this had been impractical anyway: 'I fear that if they have no other means of subsistence, they will be dead before the orders could reach the village officers.'[189]

The Government of Bombay was nonetheless stung by the perceived challenge from the native poor, and aware that such numbers leaving relief were likely to increase famine mortality, for which they were ultimately accountable. In their determination to deny responsibility for those choosing not to take relief, they sought approval for their actions from the Government of India, which supportively replied that they

> entirely share opinion of Government of Bombay as to mischief of yielding to strike on relief works attributable to combination and thinks that relief wages should not be raised if Government of Bombay are quite satisfied that the rates are sufficient to keep people alive – a point which must necessarily be left to its judgement ... The Viceroy has received telegrams from Sholapore complaining of reduction of wages but has taken no notice of them.[190]

The Bombay Government then published this exchange, allowing them to be seen responding to local protests from both workers and assistant collectors, yet to maintain their policy without risk of subsequent criticism from above. Not only did this fail, thanks to the coda that the exact wage remained Bombay's decision, it enraged Temple, who was still to take up his post as Governor and attacked his predecessor, Wodehouse, for lack of 'loyalty and imperial discipline'.[191] Temple's anger at the 'strikers' was even greater. His particularly unfortunate condemnation of 'the infatuation of these poor people in respect to eating the bread of idleness'[192] revealed a relative lack of concern at the threat to their lives, implying that this was itself a strategy: 'They counted somewhat on exciting the compassion of the authorities, and still more on arousing fears lest some accidents to human life should occur. They wandered about in bands and crowds seeking for sympathy.'[193] It is more likely that they sought food.

Though the works 'strike' muddied the waters, this concern with 'gangs' of famine-stricken people wandering throughout the country was constant during all famines in colonial India. The emergence of indigent wanderers was identified, then as now, as a key warning sign of famine. For the British, it was also seen as an indicator of social breakdown and thus of danger, necessitating urgent relief measures for the sake of state control as much as saving of life. This was set against optimistic views of the moral economy of 'village communities'. The 1867 Famine Commission Report asserted that, if left alone, villages' 'wonderful social poor law' would keep people alive and, at the same time, preserve the conservative social structure over which the British could retain hegemony.[194] By 1901 the theme was unchanged, the commission arguing that above all relief should not 'impair the structure of society, which, resting as it does in India upon the moral obligation of mutual assistance, is admirably adapted for common effort against a common misfortune'.[195]

This was particularly sanguine in a poor district in which the land revenue system had deliberately fragmented villages and weakened headmen in the desire to create literally 'independent' *ryots*.[196] Nonetheless, wanderers who neither took relief nor stayed at home threatened colonial logic as well as the safety of the public and the success of the campaign. Ashburner's pejorative exasperation was typical: 'The perversity of the people in this respect would be incredible to those who had not experienced it. They would leave food and shelter in the poor-houses for the certainty of starvation outside.'[197] However, differential wages and the distance test themselves required many workers to leave their communities, while harsh relief conditions inevitably exacerbated their tendency to seek out alternative means of survival. British concern was less with the implication of inadequate relief, or its consequences, than with public order. Relief Overseer Raghunath Ramchandra, for example, warned his superiors 'I shall have to turn out all such men who refuse to perform the task-work allotted to them but ... it is probable that these people would disturb the country if they are turned out.'[198] Stanley Wolpert suggests that many who left their homes during the famine indeed joined semi-political criminal bands.[199]

There was therefore a desire to control and discipline the famished population, strengthened by their unwillingness to take relief. Hewlett prompted a lengthy debate by suggesting that all famine victims, including those in villages, should be rounded up: 'I certainly would not allow famine-stricken people to wander about the country. I would compel them to go to relief works or poor-houses. If left alone, they wander till they die. By crowding into a town they disseminate disease.'[200] Though this proposal was rejected as administratively, economically and politically impractical, it struck a chord with many, showing up another tension in the non-interventionist paradigm. Despite restricting access to works by using market principles, the British found it hard to accept the right of those who were eligible for relief to choose alternative survival strategies.

The threat to social order was also seen, like famine itself, as a moral breakdown, and it was assumed that many would respond to the crisis by resorting to crime. The figures for Ahmednagar in Table 5.6 show that

Table 5.6 Offences and convictions, Ahmednagar district, 1874–78

Year	Offences	Convictions
1874–75	4,076	2,371
1875–76	3,229	2,596
1876–77	2,628	2,023
1877–78	3,775	4,319

Sources: Boswell to Havelock, No. 1952, 20 July 1876, p. 256; GOB Resolution No. 4280, 23 August 1878, p. 578.

this was not especially true. Though there were over a thousand more theft cases in 1877–78,[201] this partly reflected low crime levels in the previous year of the famine. Most additional crime came in the form of petty stealing and a few large grain robberies, rather than violence.[202] While David Arnold has highlighted attacks on grain stores and individuals as a form of protest during the famine in Madras,[203] it would appear that few in Ahmednagar risked such actions after the Deccan riots. The authorities also remained severe on all crime. Indeed, Table 5.6 shows the rate of convictions accelerating faster than – and even, bizarrely, overtaking – that of offences during the riot and famine periods.

A common problem during famines in British India was for prisons – with free food – to become more attractive than relief works. According to *Indu Prakash*, this was true in Bombay Presidency even before the reduction of Civil Agency wages.[204] With convictions increasing, Ahmednagar jail was reported to be overcrowded by October 1877, but inmates were not to get subsistence more easily this time: 450 of them were transferred to the railway works.[205] Apart from being guarded, their conditions were similar to other famine labourers, though they benefited from being fed rather than paid.

A more common strategy for peasants was migration. As early as September 1876, Jacomb reported that 'whole villages have been and are emigrating'.[206] Jenkin Jones argued that whereas agricultural labourers were likely to require considerable relief, 'the Kunbies who have many of them departed with their cattle towards Nasick or the Western districts of Khandeish' would not.[207] It was on a similar basis that Temple drew his distinction between 'enterprising' *ryots* and 'idle' wanderers, but the government remained ambivalent to peasant departures. As seen in Chapter 4, the land revenue system was harsh on those attempting to evade payments in bad years and then return. A few poorer peasants abandoning their holdings may have been acceptable, but large scale semi-permanent migration would be disastrous for agricultural recovery and reflect badly on the government – especially as many migrants headed not for northern British districts but the *Nizam* of Hyderabad's Dominions to the east.

According to the *Nizam*'s famine report, 'Under the influence of the first scare and the pressure caused by want, the inhabitants of the British territories became panic-stricken, and were reported to be emigrating en masse for H. H. the Nizam's dominions ... which they probably considered a land of plenty for themselves and their cattle.'[208] The *Nizam*'s ministers made great play of allowing those who did 'to receive relief without any distinction, and exactly in the same manner as His Highness's own subjects'.[209] This was partly to claim moral high ground over the Government of Madras, which refused relief to the *Nizam*'s subjects, but also to take the logic of *laissez-faire* – which the British had persuaded the *Nizam* to follow – to its logical conclusion: 'in such critical times as those of widespread famine, raiyats should be allowed perfect freedom to proceed without restraint wherever they would, with the view to their own convenience in securing a livelihood'.[210] Nor was such competitive liberality confined to government. Just over the border from Ahmednagar, it was claimed, 'The people of Ashte treated their immigrant brethren with remarkable kindness. They allowed many of them to take shelter in their houses, gave protection to their cattle, and employed such as were willing to labour in their fields ... The poor know how to feel for the poor.'[211] Given that the famine did not stop at the border, this was a sanguine view and the *Nizam*'s Relief Committee was pressed by Temple to concede that immigration from Sholapur and Ahmednagar had been exaggerated.[212] Unlike the British, however, they kept detailed records of migration in both directions, given in Table 5.7.

Though these figures should be treated with caution, they suggest significant ex-migration from Ahmednagar. British subjects numbering 32,553 are shown arriving in parts of the *Nizam*'s Dominions bordering the district, and a further 8,991 possibly coming from the same direction. The combined total constituted over 5 per cent of Ahmednagar's population and 88 per cent of all emigration to the *Nizam*'s Dominions. By contrast, migration between the *Nizam*'s Dominions and Madras Presidency tended to go in the opposite direction. Temple's desire to play down the extent of emigration therefore implied his lack of interest in the particular problems of the isolated Ahmednagar population, as well as irritation with the princely state. This was also reflected in the Government of Bombay's response to a proposal by the *Nizam*'s British

Table 5.7 Migration between British India and the *Nizam*'s Dominions during the 1876–78 famine

Nizam's district	From *Nizam*'s to British				From British to *Nizam*'s			
	Men	Women	Children	Total	Men	Women	Children	Total
Raichur*†	1,052	861	761	2,674	101	75	42	218
Parbhani‡	—	—	—	—	3,281	2,642	—	5,923
Bidar§	—	—	—	—	919	791	568	2,278
Nander‡	—	—	—	—	1,490	852	726	3,068
Naldurg*§	(not disaggregated)			486	(not disaggregated)			7,944
Nalgunda*†	7	—	—	7	13	—	—	13
Shorapur*	(not disaggregated)			17	43	36	28	107
Gulbarga*§	1,371	573	3,112	5,056	1,230	1,283	551	3,064
Lingsugur*†	2,532	2,122	2,356	7,010	64	55	56	175
Aurangabad¶	(not disaggregated)			186	4,927	3,251	1,703	9,881
Birh*¶	—	—	—	—	6,544	4,556	3,628	14,728
Total	4,962	3,556	6,229	15,436	18,612	13,541	7,302	47,399

* Famine affected districts; ¶ Districts bordering Ahmednagar; † Districts bordering Madras Presidency; ‡ Districts a short way inland from Ahmednagar; § Districts bordering Sholapur.

Source: *Nizam*'s Famine Report, p. 148.

executive engineer, Bennett Fitch, to co-operate in building a road through the serrated border of Jamkhed and Ashti *talukas*. Though this would have benefited trade between southern Ahmednagar district and the railway, it was rejected because the work was not needed for relief.[213]

The reluctance to acknowledge the extent of voluntary emigration from Ahmednagar was curious given the Government of Bombay's proposal, late in 1876, to assist migration from the district to the Central Provinces.[214] This was rejected when both Temple and Lytton pointed out that the real problem of the Deccan was underpopulation,[215] and also because few were willing to leave permanently.[216] Though many took the risks inherent in migration, *ryots* – as opposed to landless labourers – expressed the intention to return unless their resources were already exhausted. Woodburn reported that 'they would rather die than desert their farms'.[217] This confirms de Waal's insight that survival strategies come second to the preservation of livelihoods.[218] Migration from Ahmednagar was usually reported as an attempt to find fodder to keep livestock alive. Table 5.7 shows that relative to famine norms a high proportion of migrants was women and children, suggesting that many families travelled together rather than disintegrating, though a significant number of women were also left behind, dependent on relief.

Even family migration remained a dangerous strategy, with fodder and clean water hard to find on the road. Migrants and their cattle were often described as emaciated. As deaths outside villages or relief camps went unregistered,[219] along with successful emigration, their final fate can only be guessed from the overall decline in the population and descriptions such as Stewart's in 1878: 'the results of the calamity are still apparent in the number of tumble-down houses to be seen in every village the occupants of which (such is the universal tale) emigrated and have never returned'.[220] Those who did come back were liable 'to find their houses in ruins, as thatch was stolen for fodder and rafters for firewood'.[221] Permanent emigration was still being reported from the district years later,[222] which Collector King called 'a sequel of the famine'.[223]

If emigration was therefore far from ideal for *ryots* and unpopular with the authorities, so was the idea of their coming onto relief camps intended for the landless. As with emigration, however, more *ryots* successfully sought relief close to home – with the apparent connivance of Ahmednagar officers – than Temple wished to admit. He maintained to the end that 'a comparatively small portion of the lesser ryots and subordinate cultivators resorted to relief works, but the mass of the ryots or peasant proprietory, who constitute the real agricultural community, never came on relief at all'.[224] There were no specific rules preventing *ryots* from taking relief, but Temple had warned against it on his first visit to Bombay on the assumption that most had food stocks,[225] and before the 1877 monsoon failed the Government of Bombay declared that an

> inestimable advantage will be lost, if those on whose labour the tillage and cultivation of the land depend should cling, in large numbers, to the charity of the State, instead of making efforts to help themselves, and to resume their independent occupations. [Therefore] all holders or sub-holders of land, who may be in good physical condition, should be relegated to their villages, and told that they should now betake themselves, as in ordinary years, to the cultivation of their fields.[226]

This recognised, however, that many *ryots* had sought relief. The Famine Commission concurred that it had been given to many 'lesser ryots and sub-tenants' but calculated that only 3 per cent of Civil Agency workers in Ahmednagar were agriculturalists, the majority being low caste or tribal *Bhils, Mhars* and *Kolis*.[227] However James Gibbs, the longest-serving member of the Government of Bombay, declared Temple 'only generally correct' on the *ryot* question, as he was 'inclined to think that many of the better class of ryots did have recourse to the relief works'.[228] Equally importantly, he pointed out that this could only be a matter for conjecture, for both the government and the Famine Commission, as the status of those on relief was not recorded. In Ahmednagar, moreover, the distinction between *ryots* and agricultural labourers was never clear. Many peasants, having reached the bottom of their downward spiral, effectively became labourers by taking relief or migrating.[229]

Those *ryots* who could find relief work locally without first surrendering their assets had a chance to maintain their livelihoods, and there is evidence that many succeeded in this. Ashburner, for example, noted that over half the workers on the Sinnar–Goti road work in Akola were cultivators, hence his recommendation that they should not be transferred to the railway.[230] Their obstacles were the distance test and Temple's reluctance to support this strategy although he had earlier praised it in Bihar. Initially, Jacomb took the same stance, ordering the rejection of the 'well off' who 'demand that work should be brought up to their very door steps'.[231] This was changed by the introduction of the new Civil Agency wages. Hamilton reported that every worker willing to accept them in his *talukas* lived nearby.[232] Even Temple, defending the rate to the Viceroy, claimed his aim had been to set 'the minimum of subsistence for persons of slender physique doing light work close to home'.[233] Meanwhile, Woodburn discovered that 'every one of the 200 people employed belonged to villages within four miles' of a work at Kadeh in Jamkhed.[234] Fforde confirmed that the non-application of the distance test was ideal for *ryots* who were 'openly avowing that they preferred lighter tasks near their homes to the greater labour and discomfort on the works under professional supervision'.[235] He calculated that accepting the lower wage would lose non-PWD workers eight rupees if they stayed on relief from November to June. This could easily enough be raised by those with bangles or utensils to sell, or access to credit, but those wholly dependent on relief, such as widows, were 'really badly off'.[236] The Government of Bombay concluded only that the Temple wage seemed adequate, suggesting that they were prepared to turn a blind eye to peasants with some means taking relief.[237]

Sensing this, the Assistant Collector most sympathetic to those on relief, Hamilton, went so far as to propose a new work in Akola onto which he predicted 'only the families of the cultivators in the neighbouring villages will come', while the men prepared for the monsoon. In doing so, he openly assumed that there was no longer a need 'to drive people far from their homes'.²³⁸ Jacomb did order that 'turning off the agricultural classes may be held in abeyance for the present', but still insisted on the distance test, which 'there was reason to believe ... had not been lately attended to'.²³⁹ Such apparently deliberate subversion of the rules by his assistant collectors was not inspired by sentimentality, but an attempt to make relief work in practice.

Robertson summed up the consensus that 'it is most possible that the Civil Agency rates may be ample for people near to their homes, yet quite insufficient for those at a distance'.²⁴⁰ Those forced to accept relief which adhered to both the Temple wage and the distance test struggled for survival to the point that it was scarcely relief at all. Tacitly allowing *ryots* to take relief while staying on their holdings, however, was an enlightened policy that may help to explain why famine mortality was relatively low in such a poor district as Ahmednagar. The distance test was retained when the Bombay Famine Code was published, but the committee set up to formulate it also argued, significantly, that 'In *abnormal* seasons of scarcity and distress, there is still greater reason that the discretion of the local officer should not be fettered by rules.'²⁴¹ Certainly the Ahmednagar case suggests that local relief decisions were often more responsive, less ideological and more effective than those made at a distance. Klein and Singh have argued that by the early twentieth century local officers were generally given far greater autonomy, increasing the effectiveness of British relief in India.²⁴²

Mortality

Temple recorded that 49,187 people died on relief camps in Bombay Presidency during the famine, and estimated that around another 100,000 deaths had been 'famine related'.²⁴³ In discounting a further 336,302 recorded deaths in the affected districts as 'normal', he argued that cholera, smallpox and some 'fever' mortality, though much higher than in previous years, was unrelated to the famine because the diseases had been present prior to 1876. In addition, the constant touring of British officers during the campaign had ensured that deaths were more reliably recorded than usual. This is contradicted by Sen's view that chaotic famine conditions led to a lower proportion of deaths being recorded, especially when those of 'wanderers' and migrants were neither registered nor recognised to be the state's responsibility.²⁴⁴ Nor, in Ahmednagar, were all those on relief. As late as November 1877, Jacomb confessed that he had 'no data [for] deaths that have occurred among the recipients who received Charity at the Relief Houses, but were not restricted in their residence in such houses or where no Medical Subordinates were in charge'.[245]

Sanitary Commissioner Hewlett broadly supported Temple in his total estimates, suggesting that famine mortality outside relief camps was between 94,472 and 216,498, depending on whether 'fever' should be counted. Tellingly, he said he had provided a range rather than a precise figure so as not to be seen to be underestimating famine deaths, although this claim was undermined by the admission that he suspected only a small deduction from the higher figure would be most accurate.[246] Both he and Temple were widely criticised for their sanguine conclusions.[247] Local newspapers were particularly upset by the cavalier assertion that so many lives could not have been saved by the state, when they had been in Bihar. Temple denied any change of policy, claiming 'There was an equal desire to save life to the utmost of the power of the Government and its officers', but the qualification was significant.[248] In September 1877, he had warned officers in Sholapur only to prevent '*avoidable* loss of life'.[249] This was not good enough for Famine Commissioners Sullivan and Caird, who insisted 'no financial excuse can be admitted to justify famine deaths'.[250]

The Famine Commission Report itself played down the significance of famine mortality, despite estimating it to be five and a quarter million.[251] Even including Madras, this was a remarkably high total, destroying Temple's reputation at a stroke. The 1901 Famine Commission Report was to recall that in 1876–78, 'famine relief was to a large extent insufficient, and to a large extent imperfectly organised ... The mortality was, in consequence, extremely great'.[252] Though economy had had to be his priority, Temple knew that high mortality figures would lead to condemnation of his management, hence his attempt to discount disease deaths. This argument overlooked the aggravating effect of relief conditions on the spread of cholera in particular, and was refuted by Cornish, who concluded, 'there can be no reasonable doubt that excess mortality was famine mortality and nothing else'.[253] The Famine Commission Report concurred:

> Death from famine is not as simple and easily recognisable a matter as was formerly supposed. The effect of chronic starvation is to induce functional morbid changes in the intestinal organs which, when they have gone to a certain length, are incurable, and which manifest themselves in symptoms that often imitate those of other diseases.[254]

Registered deaths were collated on a monthly basis. The figures for Ahmednagar in 1876 and 1877 are given in Table 5.8.

Table 5.8 Monthly statement of registered deaths, Ahmednagar district, 1876–77

Months	Years		Months	Years	
	1876	1877		1876	1877
January	1,067	1,682	July	2,190	3,386
February	1,052	1,524	August	3,223	3,302
March	990	2,192	September	2,972	3,701
April	1,144	2,138	October	1,900	3,325
May	1,287	2,385	November	1,800	2,699
June	1,281	2,527	December	1,734	2,030
			Total	20,640	30,891

Sources: GRABP (1877), p. cclxxi; GRABP (1878), p. ccxlvii.

Even allowing for underestimates and epidemics, it can be seen that the famine had an almost immediate impact in July 1876. Relief and winter seem to have brought some respite, though this may also have been when the largest number of deaths went unrecorded. It is striking that death rates were consistently higher in the second year of the famine than the first, and only starting to decline by its end. By contrast, Temple reported pleasure, in his final minute on the famine campaign – completed on Christmas Eve 1877 – that numbers on relief had stayed low despite a second monsoon failure.[255] Notwithstanding worsening relief conditions, this could not have reflected demand unless these statistics are interpreted literally to suggest that as more died, fewer were left to take relief. Supply, rather, was progressively curtailed, despite sanction being requested for new works as late as September. In November, under pressure from the PWD's Accounts Examiner, Jacomb reported that he had 'directed all Relief works which are still in progress in this Collectorate to be closed at the end of the current month'.[256] Thus, to the end, relief was less responsive to need than to the treasury.

The breakdown of registered deaths, given in Table 5.9, predictably shows that the highest numbers were among infants, children and the old. There were about 20 per cent more male than female fatalities – a lower proportion than in some famines, perhaps reflecting families staying together to migrate or take relief, though it is curious that this applies even to infants.

Table 5.10 shows total mortality for 1877 per thousand of population for all famine affected districts of Bombay Presidency. By this measure, Ahmednagar appears to have fared well at the height of the famine,

Table 5.9 Mortality by age group and gender, Ahmednagar district, 1876–77

Year	Male infants	Female infants	Boys	Girls	Men	Women	Male 'old'	Female 'old'	All males	All females
1876	2,636	2,067	3,198	2,879	2,281	2,237	2,893	2,449	11,008	9,632
1877	2,588	2,207	4,347	3,835	4,924	3,626	5,348	4,016	17,207	13,684

Sources: GRABP (1877), p. cclxxiv; GRABP (1878), p. ccxlix.

Table 5.10 Mortality per thousand of population, Bombay Presidency, 1877

District	Deaths per thousand	Previous 5-year average death rate (villages only)
Ahmednagar	39.94	22.51
Belgaum	74.83	22.43
Bombay city	51.55	—
Dharwar	84.87	26.52
Kaladgi	101.77	29.99
Kanara	54.12	28.73
Nasik	35.59	21.34
Poona	33.96	21.83
Satara	49.04	21.86
Sholapur	48.81	19.81

Source: GRABP (1878), p. ccxlvi.

particularly by comparison with the cotton-rich southern districts of Dharwar, Belgaum and Kaladgi, which had higher population density. With only neighbouring Nasik and Poona registering lower mortality rates, this partly reflected the geography of the drought, which was more severe further south, including in Madras Presidency. The important outlet of migration was also less available to those in southern districts. Ahmednagar famine sufferers benefited from the presence of the major Dhond–Manmad railway project, and from the sympathy of local officers with regard to the distance test. Perhaps they were well prepared for hardship too, making the famine seem less extreme than it must have done in wealthier districts, though it accelerated a long downward slide in Ahmednagar. This can be seen more clearly from the annual compilations of village census returns than from the snapshot above. Mortality was also registered yearly, but the total population statistics

Table 5.11 Rural population of Ahmednagar (village returns), 1872–81

Year to July	Rural population	Change
1872	734,574	—
1873	748,515	+13,941
1874	755,244	+6,729
1875	782,489	+27,245
1876	806,878	+24,389
1877	710,747	−96,131
1878	683,748	−26,999
1879	672,521	−11,227
1880	672,987	+466
1881	709,776	+36,789

Sources: Boswell to Havelock, No. 1645, 20 July 1874, p. 114; Boswell to Havelock, No. 2132, 20 July 1875, p. 437; Boswell to Havelock, No. 1952, 20 July 1876, p. 220; Jacomb to Robertson, No. A/4960, 19 July 1877, p. 10; Stewart to Robertson, No. 3195, 22–4 July 1878, p. 279; King to Robertson, No. 3140, 19–23 July 1879, p. 10; King to Robertson, No. 4161, 20 July 1880, p. 11; King to Robertson, No. 4584, 22–5 July 1881, p. 27; Elphinston to Robertson, No. 5730, 20 July 1882, p. 43.

are more revealing, as they include the effects of migration, unrecorded deaths and declining birth-rates. These are given in Table 5.11 for a decade surrounding the famine.

These figures, in combination, confirm that famine mortality was far higher than admitted by Temple and, additionally, suggest emigration from Ahmednagar was even greater than appeared from the *Nizam*'s statistics. This is corroborated by the similar figures for declining numbers of livestock seen in Chapter 1. Both mortality and emigration can be observed well after the end of relief. The upturn in 1881 surprised Collector Elphinston, who suggested that the timing of the census meant that many people were still waiting in vain for a good monsoon when the data was collected. Had it been later, he reckoned, 'a very great decrease would have been discovered' as emigration continued.[257]

Conclusion

It was argued in Chapter 2 that poor peasants tended, reasonably, to be risk-averse rather than progressive in managing their land and household economies. When conducting a famine relief campaign, the state was expenditure-averse in managing its economy too, but not risk-averse in trying to anticipate or avert the horrors of famine and its consequences. The concerns repeatedly emphasised in colonial reports and correspondence are those of efficiency and economy, rather than effectiveness. It has been suggested that one way to prevent economic collapse without overstretching budgets would be to offer generous blanket relief for a short period to prevent panic setting in, encouraging populations and markets to respond positively to each other.[258] The Government of Bombay's relief, by contrast, was mean, reactive and late. With a model campaign in Bihar in 1874 to draw on, and its leader in charge, the only lesson they took was the negative one that such expense should be avoided. Despite sanctimonious declarations that experience was more valuable than sentiment, the Bombay campaign was run according to pre-determined principles of reliance on the market to supply food and avoidance of *ryots* becoming dependent on relief. Grain traders were ill-equipped to supply a poor remote district like Ahmednagar, and were given a licence to exploit by the sliding scale and the refusal to allow local officers to intervene, or even keep a check on food quality. Yet adherence to classical liberal nostrums was more dogmatic than ever with so much at stake. Similarly, the tight restrictions on budgets led to relief tests and conditions so extreme that they weakened the population, through malnourishment and disease. Many chose alternatives to relief, including migration, and many needlessly died. Even in the short term, the benefits to the state of a quick recovery were sacrificed to cost-cutting, with resultant losses through further immiseration and migration lasting for years.

In an attempt to justify relief so meagre that tens of thousands saw no benefit in taking it, the government sought to blame peasants for their own response to the famine, and to discipline them. It was frequently said that more generous relief would 'demoralise' the population, by disinclining them to work elsewhere. Yet the distance test, where applied, prevented peasants from continuing on their holdings. Robertson believed that in spite of this 'the patient manner in which the people bore up against this adversity is beyond praise',[259] and even Temple admitted afterwards that this fear had been unfounded.[260] Those who were so demoralised by low relief wages that they left, however, were castigated as lazy and undeserving, Temple conveniently absolving the state of 'the responsibility of preserving every one from the consequences of his own folly or misconduct'.[261] As at other times, such labels were most readily attached to tribal groups such as the *Bhils* 'who prefer idleness to work and are generally ready to thieve',[262] but relief managers were consistently exhorted to sack 'incorrigibly idle' workers throughout.[263] Curiously, this was scarcely ever reported to have happened in Ahmednagar. Yet such language encouraged physical and financial abuse of those on relief, which was never properly addressed because of inter-departmental rivalry. Peasants were usually excluded from these charges of idleness, but in the process efforts were made to exclude them from all relief. This was blind to the numbers temporarily and permanently abandoning their holdings as a survival strategy. Moreover, the distinction between needy but lazy labourers and independent peasants was hollow, particularly in the light of the aggressive assertions seen in the previous chapter that poor landholders would be better off in the labour market.

The Bombay Government's famine campaign was characterised above all by attempts to maintain control during the crisis. As well as budgets and the famished population, they sought to keep a tight rein on their own subordinates, both native and British, through the implementation of uniform rules and the demand that every decision should be sent up for approval. This inappropriate top-down approach ensured that *ad hoc* decisions, made at a physical and emotional distance from the famine field, were informed less by emerging practical realities than by fixed paradigms and hierarchical power games. Challenges from below on the effectiveness of policies such as the disastrous Temple Wage were neutralised by re-interpretation, fear of contradiction by superiors and the greater importance given to procedure. Though the Famine Commission did request – and receive – more frank and critical appraisals of the campaign, its principles were retained and entrenched in the 1885 Famine Code. Individual agency was discouraged to the point that committed local officers like Jacomb and Hamilton could only influence the success of their relief efforts in breach of their orders.

In these circumstances it was no irony that the Bombay campaign as a whole came to be regarded as a failure. Its only success was in keeping to a stringent budget, through meanness and suspicious disbursement of costs, which decimated the resources of the population and Local Funds. A similar attempt to discount the majority of famine deaths was rejected in the Famine Commission Report. By contrast, there were limited successes in Ahmednagar that helped to keep mortality down. These related to the liberal instincts of Jacomb in opening works and feeding migrants according to demand as opposed to sanction, for which he was repeatedly condemned, and the subversion of the distance test to make a virtue of the Temple Wage for *ryots* staying on their land. These lessons were not – perhaps could not have been – learned. The Bombay Famine Code rightly raised the minimum wage, but did not abandon the distance test, which encouraged wandering that the British hated and discouraged peasant perseverance they claimed to exhort.

Ahmednagar officers' opinions and reactions to events varied and by no means did all favour the peasantry, but they recognised that the nature of relief works necessarily altered their relationship with the population. The Government of Bombay also saw relief as exceptional, refusing to recognise the accentuation of long-term problems for *ryots*. They attempted, as far as possible, to respond to famine on the same non-interventionist yet rule-bound principles that characterised their ordinary administration. In this context, the need to provide relief at all threw them into confusion over its aims, costs and utility. The result was inter-departmental tension, failure to respond to demand or need and tests so strict that – where they were applied – relief was worth next to nothing to Ahmednagar *ryots*.

Conclusion

This book has attempted to reassert the centrality of state policy in assessment of the origins of poverty, vulnerability, unrest and famine among supposedly productive peasants in colonial western India. Analysis of the impact of state legislation, investment, taxation, welfare measures and relief – or lack of them – on agrarian society in Ahmednagar between 1870 and 1884 suggests that close involvement with the state was not to the population's advantage. Indeed the closer the contact between the two, the greater the potential for harm. The Bombay Government was insensitive to the unreliability of peasant land as a factor of production, and to the lack of either market or labour opportunities, and failed to invest in infrastructural support for any of them. Early state interventions in credit markets exacerbated exploitation and the response to the Deccan Riots did more to undermine security than to redress the situation. However, the most critical factor of production, capital – which was always scarce – was expropriated by the moneylenders and also by the state itself, through a tax system which many within it believed to be unreasonable and inflexible. Thus, the gradual slide into famine can be observed more clearly among small peasants than landless labourers, for whom the crisis was more acute. In the terms set out in the introduction, poor Ahmednagar peasants were chronically food insecure because their district was remote, arid and under-capitalised. Their vulnerability was increased by falling prices for their only commodities, *jowar* and *bajri*, by reduced availability of credit and by the prospect of severe tax rises. Their ability to cope was undermined by mortgage foreclosures, restrictions on mobility, barriers to relief and the government's refusal to remit the revenue demand. Moreover, their recovery was threatened by the fact that every one of these factors continued to affect them as much, if not more, after the famine crisis as before it.

What can conclusions from Ahmednagar, a more than usually vulnerable district, at a particular historical time, contribute to broader understanding about the process of famine generally? The preceding analysis suggests that it can be characterised in three ways. First, as a continuous, long-term, downward spiral – in which peasants' defences were weakened by cumulative forces, including inflexible land revenue policy, depressed markets, hostile credit relations and declining resources. Second, emphasis can be put upon the question of risk – and especially on cultivators' failed attempts to avoid or manage it. Given that rainfall patterns in such a harsh environment made agriculture inherently risky in Ahmednagar, additional gambles on volatile commodity markets were undesirable. The combined fluctuations of markets and weather made it predictable that *ryots* would suffer if both were poor at the same time. It is therefore reasonable to conclude that they would not have taken risks on the market if they had not been forced to, by land revenue rates set with a blinkered eye only for the potential opportunities created by agricultural commercialisation. Finally, famine can be understood as a series of failures by the state to respond to the needs of the population. Its reliance on the market was driven by ideology which contradicted humanitarian impulses, especially during the crisis but also before and after it. Had markets worked perfectly, this could have been justified, but it was inevitable that they would not in a poor, remote district. The state's inability to recognise or respond to glaring market failures throughout the period of study was critical, and reflected its rigid top-down decision-making structure.

These characterisations may assist in applications of Sen's entitlements theory to other famines involving smallholders, in colonial India or elsewhere. Peasants' capacity to control exchange relations – as well as their assets – were undermined by chronic immiseration, excessive risk and lack of reasonable or expected levels of support. These were not merely economic processes – although the Government of Bombay wanted to see them as such – but ones in which state policies and philosophies of governance played an important part, whether intentionally or not. Even where states are weak, therefore, it is important to consider their roles, over time, in eroding entitlements. The relationship between the state and the population will inevitably vary in different regions and periods, but famine processes are likely to have some common features. The trade-off between market-based development in remote rural areas and smallholders' autonomy and vulnerability remains pertinent today.

Some of the factors that increased peasant vulnerability to famine in Ahmednagar in the 1870s were specific. The region's main cash crop, cotton, could only be grown in insignificant amounts. The local economy lacked diversity, reducing non-farm employment opportunities, and was poorly integrated with neighbouring markets, largely thanks to inadequate transport facilities. The peasantry itself was also unusually undifferentiated, with very little capital resting in the hands of large cultivators. These findings serve only to confirm the well-established understanding that any area wholly dependent on rain-fed foodgrain production is liable to be food insecure. In considering how a single district case study can contribute to wider debate, however, a further specific negative experience for Ahmednagar in the 1870s and 1880s stands out.

The timing of 30-year land revenue revisions immediately before and after the 1876–78 crisis created an economic shock at a time of extreme vulnerability. Anticipation of its impact reduced confidence in *ryots'* future prospects, affecting all markets, including those in land and credit. This fatally undermined security of tenure and reduced the extent of cultivation. Peasants rightly perceived painstaking enquiries into questions of land ownership and tillage as a threat and thus the state as a competitor for limited resources, rather than a guarantor against destitution. The implementation of the majority of the revisions between 1878 and 1884 also served to prolong *ryots'* chronic difficulties well beyond the recognised famine period. This was exacerbated by the disqualification of that period from the eventual generous ruling that *pot kharab* land should not be reclassified as taxable, while the initial reduction in the levy of new rates, ostensibly to allow recovery, was pitifully late, short and weak. The continued ban on revenue remission until 1881 and, worse, especially rigorous collection of suspended arrears from the years of drought further fuelled the famine process.

Famines are never caused by single factors and it is not the intention here to highlight one at the expense of others. Rather, it is argued that the significance of chronic indebtedness for peasant vulnerability has been somewhat over-emphasised by historians, as it was in the Deccan Riots and Famine Commission Reports.[1] In contrast, the land revenue system, though criticised, has at times been exonerated from direct blame for famine.[2] The evidence and analysis presented in this book suggest that that balance should be redressed. The rhythms of peasants' incremental decline – and indeed the fiercest colonial disputes and peasants' own protests – related to tax issues as much as to the mismanagement of the credit market. Indebtedness was a serious problem, worsened by the unbalancing effects of the introduction of legal jurisdiction, which undermined *ryot* security and autonomy. However, the withdrawal of that credit did more to generate the famine crisis. Again the Government of Bombay was partially culpable, with its revenue rises as well as the ill-conceived and too widely anticipated Deccan Agriculturists' Relief Act. The rigidity of the land revenue demand, in the face of both long- and short-term falls in income from cultivation, similarly put increasing pressure on peasants throughout the period, intensifying from 1875. The key difference between two closely inter-related factors from the point of view of this study is that the land revenue necessarily reveals more about the relationship of the state to peasants. The state itself was the beneficiary of any harm done to *ryots* through taxation, no matter how unintentionally, whereas bad credit legislation facilitated usury by a third party disliked as much by the government as by the debtors themselves. It was hardly surprising, therefore, that the Famine Commission should lay so much blame on *Marwaris*, although their exploitation itself was so symbiotic with the revenue system that there is no logic in attacking one and not the other.

The widespread South Indian famine crisis of 1876–78 represented an economic disaster for both peasants and the state. The ability to cope depended on having greater means than either possessed at a time when each's financial prospects had been dramatically eroded by exogenous economic and political forces. Neither coped well, but poor *ryots* whose lives had revolved around the anticipation of food insecurity fared somewhat better than their peers in more fertile districts. With the rain failing, though not to an exponentially worse degree than in some other recent and subsequent years, and with both local and global markets receding, the Government of Bombay proved itself insufficiently flexible to manage the difficulties they faced. They were also complacently unprepared for the eventuality of a large food crisis, although in the event their *ad hoc* initiatives in the early months of the famine proved less harmful to peasant welfare than the more rigorous strictures dictated from above and introduced, experimentally, by Sir Richard Temple. It is significant that Temple, despite more than meeting his fiscal objectives, was subsequently held to account for failing to prevent excess mortality. The period of study also directly prompted the Famine Codes, ensuring that relief strategies were permanently in place from then on. This was not matched, however, by the creation of reserves to finance relief. Though the famine campaigns of the 1870s, including Bihar and Madras, had cost the state £16 million, Sir John Strachey's subsequent allocation of £1.5 million *per annum* as a famine insurance fund was not cumulative. If no relief was required in a given year, it was used to reduce the national debt,[3] much as peasants did not save surpluses in good years to pay the revenue demand in bad but handed them over to *sowcars*.

As Temple argued at the time, it would have been better, from the point of view of financial as well as political management, to invest the money in agricultural and infrastructural improvements to reduce the risk of famine. However, the famine-prone regions of British India were vast and the prophylactic effectiveness of such public works as existed in Ahmednagar questionable. Rather, prevention required flexibility, anticipation and greater sympathy with the risk-minimising strategies of the peasantry. It is people who suffer from famines, not regions. The expense and inefficiency of relief is in large part due to populations' inability to help themselves. This is not because of their idle preference for subsidy – as Temple often implied but subsequently refuted – but because of the erosion of their entitlements before they were reduced to accepting it.[4] Both the Famine Commission Report and the 1885 Bombay Famine Code focused on administrative techniques of famine mitigation once it had struck, but famine is never wholly unexpected, nor a technical problem awaiting a scientific solution.[5]

Even in the 1870s, plenty was understood by local officers about famine process and the possibility of reversing it. The Government of Bombay was unwilling to engage with such perspectives, sticking religiously to its positivist agenda. If, as Sumit Guha suggests, vulnerability was caused by over-expansion of the Deccan economy in the 1860s, it was remarkable that the state should still be pushing the same modernising prescriptions on Ahmednagar *ryots*, who bore more similarity to the landless than to capitalist yeomen, during the 1870s downturn. Both its consistent lack of concern with peasant viability throughout the period and its stringent relief campaign reflected not only a lack of money but also a desire to control and discipline what the state saw as a frustratingly passive population. Understanding the famine, despite contrary local reports, as an unpredictable event bearing no relation to existing rural policy, allowed the government to treat the relief campaign as exceptional. In practice, however, relief represented the contraction of normal rules more than their contradiction. The principles of *laissez-faire* were applied more strictly than ever. Though the creation of works was an unwanted intervention, the government remained dogmatic to the point of farce in refusing to get involved in the food supply to them. They also steadfastly insisted that peasants could be supported by village communities that had long been atomised by the individualisation of the land revenue system, trying unsuccessfully to keep them off the works to prove it.

Far from attempting to respond to people's specific needs and strategies during the crisis, the state withdrew and relied on half-baked principles for guidance. Though the colonial state's capacity for economic management was limited, it was capable of doing considerably more to reduce the likelihood and impact of famine crises. That it did not reflected the conflict of such an agenda with its own fiscal imperatives. The over-reliance on the land revenue meant that, while it was not necessarily excessive *per se*, the Survey and Revenue Departments were excessively reluctant to allow for variations in the weather or prices, or for evidence of poverty. Lack of investment in cultivation, even in the form of loans, represented a similar obsession with annually balanced budgets. Above all, Temple's desperation to restrict spending in the famine years meant that rural livelihoods, as well as thousands of lives, were lost through a lack of basic support.

After the works had closed, mean attitudes towards cultivators continued at the provincial level, despite the concern of the Government of India. Not only were revenue collections increased, but the Bombay Government also contradicted its own evidence of lasting misery by contrasting it only with the famine period rather than previous years. Poor harvests, high food prices and low agricultural wages had always been seen as indicators of a lack of well-being in Annual Administration Reports, and were precisely the target of the relief works. Yet, in 1879, when the railway in Ahmednagar had been completed and wages were lower and prices almost as high as in 1877, there was no question of further assistance. Time limitations on relief, which Temple had sought to end in December 1877, before cultivation had resumed or prices fallen, were critical to the minimisation of its costs and also its effectiveness as a platform for recovery. It is no surprise that N. S. Jodha identifies the success of post-colonial western India in countering the threat of famine with the permanence of the Employment Guarantee Scheme, which in some respects resembles the temporary famine works of the colonial era.[6] Its recent underfunding may come to be regretted.

The period between 1870 and 1884 was one of long struggle for Ahmednagar peasants. The famine process can be observed at all stages. *Ryots* were forced to depart permanently from their holdings throughout the period, by a long run of poor seasons, decreasing foodgrain prices, reduced credit opportunities, increased foreclosures of mortgages, land revenue hikes and by their own inability to cope with the famine crisis in those circumstances. It is important to note that the slide into vulnerability was under way by the start of the period and was continuing – to a greater extent – at its end. The 1876–78 famine was not separate from this process but part of it. While it triggered larger numbers of departures, it did not make a significant qualitative difference to the misery of peasant experience, especially as cultivators were in some cases able to rely on relief from a state which rarely gave them any support at other times. Few who managed to retain their holdings lost their lives. Those who lost land during the crisis became much more vulnerable, but no more so than those who lost land at other times. The evolution of more commercial but less available commodity and credit markets in the early 1870s and 1880s, exacerbated by insensitive tax rises in the latter period, increased risk and put severe pressure on *ryots* that made the crisis neither unexpected nor especially extreme.

The sense of inevitability about this process of famine may in part reflect the choice of district and period, which was selected as extreme rather than representative. McAlpin is justified in chastising the nationalist Romesh Dutt's attempt to condemn the whole economic basis of imperial rule from the standpoint of famine alone.[7] However, her own argument, effectively that 1870–84 was the worst part of a longer-term period in which famine vulnerability was generally reduced, attempts equally unjustifiably to do the reverse, downplaying the significance of famine on the grounds that British policy improved the rural economy over time.[8] Her vindication of the land revenue demand, for example, shows only that it took a lower proportion of average income in 1886, after being revised, than it had before Wingate's original survey, when Pringle's rates had been exceptionally high.[9] Her own data shows the 1880s revenue revision in Nagar *taluka* to have increased the demand by 13 per cent in real terms from the start of Wingate's settlement, and 30 per cent from its end.[10] Moreover, no account is taken of the increase as an economic shock at a critical moment, which played a part in triggering famine crisis, nor of revenue payers with below average incomes. Similarly, the argument that reforms at this time to the revenue, relief and *takavi* loan systems inevitably took time to show their benefits is questionable because their effects were either short term or annual, and famine was no less devastating in Bombay Presidency in 1899 and 1901. McAlpin's agenda precisely reflects that of the Bombay Government of the day, prioritising the hope of long-term development over the management of short-term needs.

The conclusion from this shorter-term case study is that many policies designed to improve long-term prospects are inappropriate, will fail on their own terms in poorer areas without adequate support and cannot reasonably be implemented without the establishment of safety nets. The extreme suffering of one generation cannot be justified by the optimistic hope of better for their grandchildren. Fifteen years are long enough. In 1875, Boswell asked, 'Are we content to see our once happy cultivators thus ruined, and have we considered what the result of the changes going on will be? If so are we prepared to meet them?'[11] The question was rhetorical, for he knew that beyond tinkering with credit legislation, the Government of Bombay would 'do next to nothing for the cultivator'.[12] Throughout the turmoil between 1875 and 1884, *ryots* continued to be ruined and the Bombay Government continued to take minimal responsibility for their welfare, for fear of harming future prospects and wasting precious resources. Their own rhetorical efforts during the famine crisis sought to argue that it was not their fault and its effects could therefore be morally justified. Yet even a non-interventionist state must support its populations at some point. Otherwise it would be better to withdraw completely from all interactions with them, including taxation and legislation affecting their movements and credit relationships.

An important theme of this examination of famine process and policy is therefore that of state accountability, exclusively internal in the colonial context. As well as investigating the costs and benefits of particular governmental agendas, this book has explored how decisions are made in the context of both the attempt to develop a backward district and the anticipation and management of famine crisis. Frequently, this was not just significant for an understanding of the discursive elaboration of a multi-layered colonial state. The denial of evidence of suffering and down-playing of junior opinion which was best placed to judge it had a direct bearing on the famine process itself. At times, such as Temple's refusal to remit land revenue during the crisis, the primary aim seemed to be to create an impression of good management to higher echelons of the state, in that case by deferring losses until after the costs of relief had been calculated. In so doing, costs were also deferred to famished peasants, whose ability to cope and recover was directly undermined. Similarly, Strachey's drive to create uniform famine policy removed the capacity to take account of complex or locally specific processes, replacing local autonomy with paradigms designed to justify suffering. The 1901 famine in Bombay Presidency saw worse mortality and disruption than 1876–78, despite considerable improvements in organisational efficiency. A case has therefore been made here for the importance of local and individual agency. Ahmednagar had many collectors and assistants in the period of study, with a variety of different sympathies, priorities and capabilities, but all were constrained by the procedures of hierarchy from making the state more responsive to peasants if they were so inclined. Ironically, the administrative chaos of the famine crisis presented an opportunity to subvert government directives by allowing *ryot* households onto Civil Agency works near their homes, which may have been very significant in supporting coping strategies. It should not be forgotten, however, that Ahmednagar assistant collectors conceived the distance test in the first place.

Autonomy was also of central importance to peasants' capacity to manage risk, if not to become wealthy. Though their poverty and indebtedness severely limited most *ryots'* agency within an unbalanced district agrarian structure from the start, this was exacerbated by the state in several ways, including sedentisation, the institution of civil courts for debt and rigid revenue demands. More importantly, peasants had virtually no leverage against the state itself, even in negotiation with assistant collectors or *mamlatdars*. Their weapons were too weak.[13] Despite the Deccan Riots and the so-called famine strike, they had little capacity to act collectively and held little threat for the state, which, tellingly, they feared. The attempt by the Bombay Government to blame the *Poona Sarvajanik Sabha* for co-ordinating both events demonstrated their neuroses. Such elite protest, especially if it was capable of harnessing mass support, would have mattered. Though they prompted a serious bout of soul-searching, neither the riots nor the rejection of inadequate relief ultimately worried the state enough to listen. Peasants' inability to cope was never fully recognised outside the district, even after the slump in landholding and cultivation acknowledged in 1879.

The 1876 Bombay Administration Report recognised that the government was not democratically – or otherwise – accountable beyond the boundaries of the state, for which they blamed the population:

> In India legislation is not called for by the popular voice, nor does general opinion indicate where its provisions are too loose or where its constraints may pinch. Not until a Government officer has discovered an abuse, is it exposed; not till pressure has worked a sore is a grievance suspected. ... If this method offers instances of bad government and even oppression, the cause is not that Government are seeking to repress a free and enlightened population, to check any expression of opinion, and to force their theories to the extinction of national spirit, but that Government can get no answer or stimulus from the people, and that rather than go on for another century, as the people have been content to do for the last ten centuries, Government take the initiative.[14]

In truth, peasants were not quite so passive. Uninterested in legislation they may have been, but plenty of government officers were made well aware of their grievances, in conversations with them, and especially by the Deccan Riots. District collectors from Boswell to Elphinston in turn used their experience in districts like Ahmednagar to expose abuses, critique many of the government's initiatives and theories and warn persistently of pressure that was always likely to work considerably more than a sore. It was the structure of the state that prevented the kind of enlightened interaction with its subjects that they purported to want. Even local officers could barely make a difference. Native newspapers and organisations were ignored. Peasants too poor to make ends meet were actively discouraged from maintaining their livelihoods, let alone complaining at their fate. Bad government and even oppression were hidden only by self-serving colonial rhetoric such as this extract itself. What cultivators did was feel their impact, at all stages of the long process of famine.

Notes

Introduction

1. See Bruce Currey, 'Coping with Complexity in Food Crisis Management', in Bruce Currey and Graeme Hugo (eds), *Famine as a Geographical Phenomenon* (Dordrecht, 1984), pp. 183–202.
2. Amartya Sen, *Poverty and Famines: An Essay on Entitlement and Deprivation* (Oxford, 1981), p. 4.
3. Amrita Rangasami, ' "Failure of Exchange Entitlements" Theory of Famine: A Response', *Economic and Political Weekly* (*EPW*), XX (41 and 42) (1985), 1747–52, 1797–1801; Alexander de Waal, *Famine That Kills: Darfur, Sudan, 1984–85* (Oxford, 1989); David Keen, *The Benefits of Famine: A Political Economy of Famine and Relief in Southwestern Sudan, 1983–1989* (Princeton, NJ, 1989); Jenny Edkins, *Whose Hunger? Concepts of Famine, Practices of Aid* (Minneapolis, MN, 2000).
4. David Hall-Matthews, 'The Historical Roots of Famine Relief Paradigms', in Helen O'Neill and John Toye (eds), *A World without Famine? New Approaches to Aid and Development* (Macmillan, 1998), pp. 107–27.
5. Famine Commission Report (FCR) (1880), part I, p. 7.
6. Sumit Guha, *The Agrarian Economy of the Bombay Deccan, 1818–1941* (Oxford, 1985).
7. Stephen Devereux, *Theories of Famine* (Hemel Hempstead, 1993), pp. 28–32.
8. See, inter alia, B. M. Bhatia, *Famines in India: A Study in some Aspects of the Economic History of India with Special Reference to Food Problem 1860–1990*, 3rd edn (Delhi, 1991); Michelle McAlpin, *Subject to Famine: Food Crises and Economic Change in Western India 1860–1920* (Princeton, NJ, 1983); H. S. Srivastava, *The History of Indian Famines and Development of Famine Policy*. (Agra, 1968); Srinivasa Ambirajan, 'Political Economy and Indian Famines', *South Asia*, 1 (2) (1971), 19–28; Ira Klein, 'When the Rains Failed: Famine, Relief and Mortality in British India', *Indian Economic and Social History Review* (*IESHR*), 21 (2) (1984), 185–214; Sanjay Sharma, *Famine, Philanthropy and the Colonial State: North India in the Early Nineteenth Century* (Oxford, 2001); Jean Drèze, 'Famine Prevention in India', in Jean Drèze and Amartya Sen (eds), *The Political Economy of Hunger, Vol. II, Famine Prevention* (Oxford, 1990), pp. 13–122; Paul Greenough, *Prosperity and Misery in Modern Bengal: The Famine of 1943–44* (Oxford, 1982); Lance Brennan, 'Government Famine Relief in Bengal, 1943', *Journal of Asian Studies*, 47 (1988), 542–67; Mohiuddin Alamgir, *Famine in South Asia* (Cambridge, MA, 1980). The most notable exception, which looks far more closely at peasant perceptions than this book, is David Arnold, 'Famine in Peasant Consciousness and Peasant Action: Madras 1876–8', in Ranajit Guha (ed.), *Subaltern Studies III* (Delhi, 1984), pp. 62–115.
9. See, for example, Tim Dyson, 'On the Demography of South Asian Famines, Part I', *Population Studies*, 45 (1991), 5–25; Greenough, *Prosperity and Misery*; Roland Lardinois, 'Famine, Epidemics and Mortality in South India: A Reappraisal

of the Demographic Crisis of 1876–1878', *EPW*, XX (11) (1985), 454–65; Amartya Sen, 'Famine Mortality: A Study of the Bengal Famine of 1943', in Eric Hobsbawm, Witold Kula, Ashok Mitra, K. N. Raj and Ignacy Sachs (eds), *Peasants in History: Essays in Honour of Daniel Thorner* (Calcutta, 1980), pp. 194–220; Elizabeth Whitcombe, 'Famine Mortality', *EPW*, XVIII (23) (1993), 1169–79; Arnold, 'Famine in Peasant Consciousness'.
10. Neil Charlesworth, 'Rich Peasants and Poor Peasants in Late Nineteenth-Century Maharashtra', in Clive Dewey and A. G. Hopkins (eds), *The Imperial Impact: Studies in the Economic History of Africa and India* (London, 1978), pp. 97–113; Ravinder Kumar, *Western India in the Nineteenth Century: A Study in the Social History of Maharastra* (London, 1968); Sumit Guha, *Agrarian Economy*; McAlpin, *Subject to Famine*; Jairus Banaji, 'Capitalist Domination and the Small Peasantry: Deccan Districts in the Late Nineteenth Century', *EPW*, XII (33 and 34) (1977), 1375–1404.
11. David Washbrook, 'The Commercialization of Agriculture in Colonial India: Production, Subsistence and Reproduction in the "Dry South", ca. 1870–1930', *Modern Asian Studies (MAS)*, 28 (1) (1994), 129–64.
12. See Peter Robb 'Introduction. Land and Society: The British "Transformation" in India' in Peter Robb (ed.), *Rural India: Land Power and Society under British Rule* (Delhi, 1983), pp. 1–23.
13. McAlpin, *Subject to Famine*.
14. Mike Davis, *Late Victorian Holocausts: El Nino Famines and the Making of the Third World* (London, 2001).
15. Devereux, *Theories of Famine*, p. 21.
16. Studies of the 1876–78 famine, including Bhatia, *Famines in India* and McAlpin, *Subject to Famine*, do address issues such as land revenue and loans which specifically affected the peasantry, but maintain the view that it was not famine-prone.
17. For a discussion of how gradual losses of assets cause livelihoods to spiral into destitution and food insecurity, see Margaret Buchanan-Smith and Susanna Davies, *Famine Early Warning and Response: The Missing Link* (London, 1995), pp. 5–6.
18. Government of Bombay (GOB) to Secretary of State for India (SOSI), No. 27, 24 December 1880; National Archive of India, New Delhi (NAI), Government of India Home, Revenue and Agriculture Department (HRAD), Revenue Branch, February 1881, 'A' proceedings 16–18.
19. H. E. Jacomb, Acting Collector and Magistrate, Ahmednagar, to E. P. Robertson, Revenue Commissioner, Central Division (of Bombay Presidency) (CD), No. A/4960, 19 July 1877, Annual Administration Reports, Ahmednagar (AARA) (1877), Maharashtra State Archives, Bombay (MSA), GOB, Revenue Department (RD), Vol. 13 of 1877, No. 1226, p. 25.
20. FCR (1880), Appendix II, p. 55.
21. FCR (1880), part II, p. 112.
22. For an excellent regional case study of how British famine relief in India was initially conceived, in the 1830s, as a counterpoint to local philanthropic traditions, see Sanjay Sharma, *Famine, Philanthropy*, pp. 135–92. On the philosophy of the Poor Laws, see Gertrude Himmelfarb, *The Idea of Poverty: England in the Early Industrial Age* (London, 1984), p. 163.

23. Keen, *Benefits*, p. 13. See Jean Drèze and Amartya Sen, *Hunger and Public Action* (Oxford, 1989), pp. 104–6 for an argument in favour of targeting.
24. Stephen Devereux, Paul Howe and Luka Biong Deng, 'Introduction', *IDS Bulletin*, 33 (4) (2002), 5.
25. *Bombay Gazette*, cited in William Digby, *The Famine Campaign in South India*, Vol. I (London, 1878), p. 370.
26. See Edward Clay and Bernard Schaffer, *Room for Manoeuvre: An Exploration of Public Policy in Agriculture and Rural Development* (London, 1984).
27. See Eric Stokes, *The English Utilitarians and India* (Oxford, 1959); Tom Metcalf, *The Aftermath of Revolt: India, 1857–1870* (Princeton, NJ, 1964); and James C. Scott, *Seeing Like a State: How Certain Schemes to Improve the Human Condition Have Failed* (New Haven, CT, 1998).
28. See Clive Dewey, *Anglo-Indian Attitudes: The Mind of the Indian Civil Service* (London, 1993); Bradford Spangenberg, *British Bureaucracy in India: Status, Policy and the ICS in the Late Nineteenth Century* (Delhi, 1975); and David Washbrook, 'Law, State and Agrarian Society in Colonial India', *MAS*, 15 (1) (1981) 649–721.
29. Robertson to GOB, No. R/3701, 9 August 1882, AARA (1882); MSA, GOB, RD Vol. 10 of 1882, No. 303, p. 308.
30. Hall-Matthews, 'Historical Roots', p. 125.
31. Sir Richard Temple to SOSI, Lord Salisbury, 22 March 1877; Oriental and India Office Collections, British Library (OIOC), Temple Collection.
32. See, for example, David Hardiman (ed.), *Peasant Resistance in India, 1858–1914* (Oxford, 1992); Hamza Alavi, 'Peasants and Revolution' in Kathleen Gough and H. P. Sharma (eds), *Imperialism and Revolution in South Asia* (New York, 1973), pp. 291–337; Eric Stokes, *The Peasant and the Raj: Studies in Agrarian Society and Peasant Rebellion in Colonial India* (Cambridge, 1978); Arnold, 'Famine in Peasant Consciousness'.
33. Ranajit Guha, 'On Some Aspects of the Historiography of Colonial India', in Ranajit Guha (ed.), *Subaltern Studies I* (Delhi, 1982), pp. 1–9.
34. See C. A. Bayly, *Indian Society and the Making of the British Empire* (Cambridge, 1988); or Neil Rabitoy, 'System v Expediency: The Reality of Land Revenue Administration in the Bombay Presidency, 1812–20', *MAS*, 9 (4) (1975), 529–46.
35. Digby, *The Famine*, Vol. I, p. 333.
36. J. King, Collector and Magistrate, Ahmednagar, to Robertson, No. 4584, 22–5 July 1881, AARA (1881), MSA, GOB, RD, Vol. 12 of 1881, No. 1743, p. 254.
37. Ian Hacking, *The Taming of Chance* (Cambridge, 1990).
38. Clive Dewey, '*Patwari* and *Chaukidar*: Subordinate Officials and the reliability of India's Agricultural Statistics', in Clive Dewey and A. G. Hopkins (eds), *Imperial Impact: Studies in the Economic History of Africa and India* (London, 1978), pp. 280–314.
39. H. B. Boswell, Collector and Magistrate, Ahmednagar, to W. H. Havelock, Revenue Commissioner, Southern Division (of Bombay Presidency) (SD), No. 1645, 20 July 1874, AARA (1874); MSA, GOB, RD, Vol. 3 of 1874, No. 1367, p. 116.
40. Boswell to Havelock, No. 1952, 20 July 1876, AARA (1876); MSA, GOB, RD, Vol. 5 of 1876, No. 752, p. 226.
41. *Ibid.*, pp. 228–9.

42. See Deccan Riots Commission Report (DRCR), Appended evidence of unnamed Sub-Judge of Ahmednagar, p. 1, and Ian Catanach, 'Agrarian Disturbances in Nineteenth Century India', in David Hardiman (ed.), *Peasant Resistance in India 1858–1914* (Delhi, 1992), 192–3.
43. Charlesworth, 'Rich Peasants'.

1 Landholding, Peasant Production and Rainfall

1. *Bombay Gazetteer, Vol. XVII: Ahmadnagar* (Poona, 1884), p. 1.
2. DRCR, pp. 13–16.
3. *Ibid.*, p. 22.
4. Colonel W. C. Anderson, Bombay Survey and Settlement Commissioner, to Robertson, No. 260, 26 February 1879; NAI, HRAD, Revenue Branch (RB), September 1879, 'A' proceedings 68–71, p. 16.
5. Christine Kinealy, *A Death-Dealing Famine: The Great Hunger in Ireland* (London, 1997), pp. 16–17.
6. FCR (1880), Part II, pp. 130–1.
7. FCR (1880), Appendix I, p. 164.
8. GOI, Public Works Department (PWD) Resolution, 16 May 1878; FCR (1880), Appendix II, p. 6.
9. General Report on the Administration of the Bombay Presidency (GRABP) (1873), pp. 118–19.
10. DRCR, p. 14.
11. Robertson to GOB, No. R/3701, 9 August 1882, p. 314.
12. FCR (1880), 'Information and Evidence collected in Bombay Presidency to answer the Questions issued by the Famine Commission', chapter 1, question 8.
13. See, for example, King to Robertson, No. 4584, 22–5 July 1881, pp. 59–68.
14. GRABP (1873), p. 34.
15. *Ahmadnagar Gazetteer*, Revised edition 1976, pp. 299–300.
16. *Ibid.*, p. 299. See also David Hardiman, *Feeding the Baniya: Peasants and Usurers in Western India* (Delhi, 1996).
17. *Ahmadnagar Gazetteer*, 1976, p. 198.
18. DRCR, p. 24.
19. See, for example, Boswell to GOB, No. 3545, 16 December 1875, AARA (1875); MSA, GOB, RD, Vol. 3 of 1875, No. 971, p. 782.
20. *Ahmadnagar Gazetteer*, p. 242.
21. See, for example, FCR (1880), 'Information and Evidence', chapter 1, question 10, answer by Robertson.
22. Quoted in DRCR, Appendix A, pp. 138–9.
23. See Washbrook, 'Law, state', 670–3.
24. Ravinder Kumar, 'The Deccan Riots of 1875', in Hardiman (ed.), *Peasant Resistance*, pp. 160–1.
25. T. S. Hamilton, Acting 2nd Assistant Collector and Magistrate, Ahmednagar to King, No. 276, 5 July 1879, AARA (1879); MSA, GOB, RD, Vol. 16 of 1879, No. 1064, pp. 159–60.
26. *Ibid.*, p. 160.
27. See Hardiman, *Feeding the Baniya*, p. 198 and Sumit Guha, 'The Land Market in Upland Maharashtra', *IESHR*, 24 (2) (1987), 121.

28. MSA, GOB, RD, Vol. 22 of 1872, No. 1395.
29. GOB Resolution, No. 3283, 10 June 1875; MSA, GOB, RD, Vol. 109 of 1875, No. 1094.
30. DRCR, p. 14.
31. *Ahmadnagar Gazetteer*, p. 245.
32. FCR (1880), Appendix I, p. 57.
33. For example, in 1876–77, Ahmednagar's 162,897 acres compared with 60,973 in Poona, 32,093 in Satara and just 22,805 acres in Sholapur district. Robertson to GOB, No. R/626, 16 February 1878, p. 11.
34. Thus the missing figure in Table 1.3 for land abandoned in 1872–73 was at least 25,000 acres. Boswell to Havelock, No. 1645, 20 July 1874, pp. 138–9. There is no extant Annual Administration Report for Ahmednagar before 1873–74.
35. Sumit Guha, *Agrarian Economy*.
36. King to Robertson, No. 4163, 20 July 1880, AARA (1880), p. 253.
37. Robertson to GOB, No. R/2431, 2 August 1880, AARA (1880), pp. 334–5.
38. Balkrishna Deora, Deputy Collector and Magistrate, Ahmednagar, to King, No. 248, 16 July 1881, AARA (1881), pp. 371–2.
39. King to Robertson, No. 4586, 22–5 July 1881, AARA (1881), pp. 427–8.
40. A. F. Woodburn, Acting 2nd Assistant Collector, Ahmednagar, to Boswell, No. 59, 5 July 1876, AARA (1876), p. 319.
41. Boswell to Oliphant, No. 2132, 20 July 1875, pp. 461–2.
42. Hamilton to Boswell, No. 7, 4 July 1874, AARA (1874), pp. 83–4.
43. Boswell to Havelock, No. 1644 of 1874, 20 July 1874, AARA (1874), p. 99.
44. GRABP (1874), pp. xxxix–xl.
45. GRABP (1875), p. 116.
46. See, for example, R. E. Candy, 1st Assistant Collector, Ahmednagar, to King, No. 561, 12 July 1880, AARA (1880), p. 196, including the complaint that even *ryots* close to Nagar city never purchased manure.
47. Ester Boserup, *The Conditions of Agricultural Growth* (London, 1965).
48. See, for example, GRABP (1874), p. xx.
49. See Srinivasa Ambirajan, 'Malthusian Population Theory and Indian Famine Policy in the Nineteenth Century', *Population Studies*, 30 (1) (1976), 5–14.
50. GRABP (1874), p. xiv.
51. Neil Charlesworth, *Peasants and Imperial Rule: Agriculture and Agrarian Society in the Bombay Presidency 1850–1935* (Cambridge, 1985), pp. 44–5.
52. GRABP (1871), p. ii.
53. Michelle McAlpin, 'Railroads, Cultivation Patterns and Foodgrains Availability: India 1860–1900', *IESHR*, 12 (1) (1975), 43–60; Peter Harnetty, 'Cotton Exports and Indian Agriculture, 1861–1870', *Economic History Review*, 24 (3) (1971), 414–29; Sumit Guha, *Agrarian Economy*, pp. 54–60.
54. W. W. Loch, Acting 2nd Assistant Collector and Magistrate, Ahmednagar, to Boswell, No. 253, 4 July 1874, AARA (1874), pp. 46–7, in which Loch argues that a 30 per cent revenue rise would leave peasants no worse off.
55. Havelock to GOB, No. 2831, 4 August 1874, AARA (1874), p. 104.
56. GRABP (1876), pp. lxxi–lxxii.
57. For example, Boswell to Havelock, No. 1645, 20 July 1874, and Stewart to Robertson, No. 3195, 22–4 July 1878, pp. 294–5.
58. FCR (1880), 'Bombay Information and Evidence', chapter 1, question 6, answer by C. S. Anderson, Assistant Collector of Ahmednagar.

59. *Ibid.*
60. Elphinston to Robertson, No. 5730, 20 July 1882, pp. 90–2.
61. DRCR, p. 14.
62. Jacomb to Robertson, No. A/4960, 19 July 1877, p. 17.
63. Boswell to Havelock, No. 1645, 20 July 1874, p. 116.
64. Boswell to Havelock, No. 1952, 20 July 1876, p. 226.
65. Stewart to Robertson, No. 3195, 22–4 July 1878, p. 285.
66. Woodburn to Jacomb, No. 481, 5 July 1877, AARA (1877), p. 169.
67. Stewart to Robertson, No. 3195, 22–4 July 1878, p. 285.
68. See, for example, King to Robertson, No. 4161, 20 July 1880, p. 16.
69. Elphinston to Robertson, No. 5730, 20 July 1882, pp. 60–1.
70. DRCR, Vol. II, p. 40.
71. See, for example, the report on Kopargaon *taluka* in Elphinston to Robertson, No. 5730, 20 July 1882, pp. 28–9.
72. See, for example, the report on Nagar *taluka* in C. G. Blathwayt, 1st Assistant Collector and Magistrate, Ahmednagar, to Boswell, No. 168, 7 July 1876, AARA (1876), pp. 348–9.
73. GRABP (1872), p. 5.
74. GRABP (1874), p. ix.
75. *Ahmadnagar Gazetteer*, pp. 284–93.
76. Lt-Gen Richard Strachey, 'Physical Causes of Indian Famines', FCR (1880), Appendix I, pp. 1–7.
77. Bhatia, *Famines in India*, p. 3.
78. GOB to all collectors, 20 November 1876; MSA, GOB, General Department, unnumbered volume of famine statements, 1878.
79. See Bombay Director of Agriculture Harold Mann, *Rainfall and Famine: A Study of Rainfall in the Bombay Deccan, 1865–1938* (Bombay, 1955), p. 4 and also Ian Catanach, *Rural Credit in Western India, 1875–1930: Rural Credit and the Co-operation Movement in the Bombay Presidency* (Berkeley, 1970), p. 6.
80. Havelock to GOB, No. 4799, 9 December 1873; MSA, GOB, RD, 1874, Vol. 11, No. 62, p. 259.
81. *Ahmadnagar Gazetteer*, p. 12.
82. Jacomb to Robertson, No. A/4960, 19 July 1877, p. 8.
83. GOB, RD, Resolution No. 5220, 25 August 1877, AARA (1877), p. 291.
84. Sir Richard Temple, 'The Famine of 1876 and 1877 in the Bombay Presidency', *Minutes by the Governor of Bombay, 1877–80*, Vol. II, OIOC, p. 92.
85. DRCR, p. 21.
86. Sen, *Poverty and Famines*.
87. GOB, 'Weekly Famine Statements', 31 October 1876.
88. Woodburn to Boswell, despatching the report of his predecessor Loch, No. 271, 5 July 1875, AARA (1875), p. 601.
89. Elphinston to Robertson, No. 5733, 20 July 1882, AARA (1882), pp. 563–4.
90. King to Robertson, No. 4584, 22–5 July 1881, AARA (1881), p. 19.
91. Woodburn to Jacomb, No. 481, 5 July 1877, p. 164.
92. Woodburn to Stewart, No. 291, 5 July 1878, AARA (1878), pp. 496–8.
93. Hamilton to Stewart, No. 514, 8 July 1878, AARA (1878), pp. 453–4.
94. Candy to King, No. 561, 12 July 1880, p. 194.
95. See, for example, King to Robertson, No. 3140, 19–23 July 1879, pp. 7–9.

228 *Notes*

96. King to Robertson, No. 4161, 20 July 1880, p. 9.
97. Candy to King, No. 561, 12 July 1880, pp. 197–8.
98. See, for example, King to Robertson, No. 4584, 22–25 July 1881, p. 22.
99. *Ibid.*, p. 23.
100. Jacomb to Robertson, No. A/4960, 19 July 1877, p. 47.
101. Woodburn to Stewart, No. 291, 5 July 1878, p. 499.
102. Hamilton to Jacomb, No. 374, 5 July 1877, AARA (1877), p. 123.
103. Jacomb to Robertson, No. 4961, 19 July 1877, AARA (1877), pp. 152–3.
104. Robertson to GOB, No. 3620, 3 August 1877, AARA (1877), pp. 261–2.
105. GOB Resolution, No. 5220, 25 August 1877, p. 291.
106. SOSI to GOB, No. 3, 6 February 1879, NAI, Government of India, Revenue, Agriculture and Commerce Department (RACD), Land Revenue and Settlement Branch (LRSB), March 1879, 'B' proceedings 35.
107. King to Robertson, No. 3141, 19–23 July 1879, AARA (1879), p. 141.
108. Hamilton to King, No. 594, 5 July 1880, AARA (1880), p. 276.
109. N. G. Deshpande, Deputy Collector, Ahmednagar, to Elphinston, No. 599, 10 July 1882, AARA (1882), pp. 521–2.
110. Robertson to GOB, No. R/3701, 9 August 1882, p. 326.
111. Woodburn to Jacomb, No. 481, 5 July 1877, pp. 171–2.
112. Hamilton to King, No. 594, 5 July 1880, p. 275.
113. King to Robertson, No. 4164, undated, AARA (1880), pp. 309–10.
114. Candy to King, No. 561, 12 July 1880, pp. 195–6.
115. Hamilton to King, No. 592, 9 July 1881, AARA (1881), p. 306; King to Robertson, No. 4586, 22–5 July 1881, p. 427 and No. 4584, 22–5 July 1881, pp. 13–18.
116. Deora to King, No. 248, 16 July 1881, p. 370.
117. Boswell to Havelock, No. 1642, 20 July 1874, AARA (1874), p. 34.
118. Elphinston to Robertson, No. 5730, 20 July 1882, pp. 85–6.
119. *Ibid.*, p. 102.
120. Robertson to GOB, No. R/3701, 9 August 1882, pp. 311–13.
121. GRABP (1878), p. iv.
122. Candy to Stewart, No. 514, 11 July 1878, AARA (1878), p. 404.
123. Boswell to Oliphant, No. 2132, 20 July 1875, p. 431.
124. *Ibid.*
125. *Ibid.*, p. 432.
126. Stewart to Robertson, No. 3195, 22–4 July 1878, p. 294.
127. King to Robertson, No. 4584, 22–5 July 1881, pp. 7–10.
128. FCR (1880), 'Bombay Information and Evidence', chapter 1, question 7, answer by Robertson.
129. Boswell to Oliphant, No. 2132, 20 July 1875, p. 494.
130. King to Robertson, No. 4161, 20 July 1880, pp. 77–8, and Robertson to GOB, No. R/2431, 2 August 1880, p. 335.
131. W. D'Oyly, Collector of Ahmednagar, to GOB, No. 2368, 1 December 1871; MSA, GOB, RD, Vol. 2 of 1873, No. 1351.
132. GOB, RD, to RACD, No. 5732-283R, 16 October 1873; *ibid.*
133. GRABP (1874), pp. 292–3.
134. FCR (1880), 'Bombay Information and Evidence', chapter 1, question 7, answer by Robertson.
135. King to Robertson, No. 4584, 22–5 July 1881, pp. 148–51.

136. King to Robertson, No. 4161, 20 July 1880, pp. 74–6.
137. Boswell to Havelock, No. 1645, 20 July 1874, p. 151.
138. FCR (1880), 'Bombay Information and Evidence', chapter 1, question 23, answer by Mr Shuttleworth, Conservator (ND).
139. See Mahesh Rangarajan, *Fencing the Forest: Conservation and Ecological Change in India's Central Provinces 1860–1914* (Delhi, 1996), p. 94.
140. Boswell to Oliphant, No. 2132, 20 July 1875, p. 478.
141. Boswell to Havelock, No. 1952, 20 July 1876, p. 242.
142. Jacomb to Robertson, No. A/4960, 19 July 1877, p. 35.
143. GOB Resolution, RD, No. 4672, 2 September 1879, AARA (1879), p. 284.
144. For example, in the *talukas* of Parner, Sangamner, Kopargaon and Akola, 136,400 acres were handed to the Forestry Department, 34,078 of which had been assessed land. Hamilton to King, No. 276, 5 July 1879, p. 161.
145. King to Robertson, No. 3140, 19–23 July 1879, p. 41.
146. Candy to King, No. 561, 12 July 1880, p. 196.
147. King to Robertson, No. 4161, 20 July 1880, p. 252.
148. Stewart to Robertson, No. 3196, 22–4 July 1878, AARA (1878), p. 445.
149. FCR (1880), 'Bombay Information and Evidence', chapter 4, question 17, answer by Colonel Anderson.
150. Loch to Boswell, No. 253, 4 July 1874, p. 59.
151. Boswell to Oliphant, No. 2132, 20 July 1875, p. 458.
152. *Ahmadnagar Gazetteer*, p. 251.
153. FCR (1880), 'Bombay Information and Evidence', chapter 1, question 4.
154. GRABP (1872), p. 310.
155. GRABP (1873), p. 438.
156. GRABP (1875), p. xlvi.
157. GRABP (1877), pp. lxi–lxii.
158. Secretary of State Lord Salisbury first proposed the Famine Commission to Strachey, 'in order to save ourselves from the irrigation quacks'. Lance Brennan, 'The Development of the Indian Famine Code' in Bruce Currey and Graeme Hugo (eds), *Famine as a Geographical Phenomenon* (Dordrecht 1984), p. 98.
159. FCR (1880), 'Bombay Information and Evidence', chapter 1, question 26, answer by Mr Propert, Collector of Khandesh.
160. Loch to Boswell, No. 253, 4 July 1874, p. 56.
161. Woodburn to Boswell, No. 59, 5 July 1876, pp. 324–6.
162. Jacomb to Robertson, No. A/4960, 19 July 1877, pp. 9–10.
163. See, for example, Boswell to Oliphant, No. 2132, 20 July 1875, p. 456.
164. NAI, GOI, RACD, Abstract of proceedings for August 1873, 21 August 1873.
165. Boswell to Havelock, No. 1952, 20 July 1876, p. 252.
166. Memorandum by A. O. Hume, Secretary, RACD, 4 February 1879; NAI, RACD, LRSB, April 1879, 'A' proceedings 6–7, Keep-With section. Original emphasis.
167. SOSI to GOB, No. 1, 15 January 1880; NAI, HRAD, RB, February 1880, 'A' proceedings 95–6.
168. GOI, PWD, Resolution No. 2271, 8 June 1883; NAI, GOI, Revenue and Agriculture Department (RAD), RB, June 1883, 'B' proceedings 99.
169. See, for example, GOI, PWD, Resolution No. 471, 1 February 1884; NAI, RAD, RB, February 1884, 'B' proceedings 32.

2 Market Opportunities, Risks and Failures

1. GRABP (1873–74), pp. cvii–cix; (1877–78), pp. cxvi–cxvii.
2. Between 1 January and 31 August 1878, 19,118 people died in the district, against a 5-year average for the same months of 11,529. Register of deaths in the Bombay Presidency; MSA, GOB, RD, Famine Branch, Vol. 115 of 1876–77, p. 177.
3. *Ahmadnagar Gazetteer*, p. 271.
4. In 1874, the GOB noted that 'The very great interest that attaches to this particular staple makes it advisable to treat of everything pertaining to its cultivation and trade with some degree of minuteness.' GRABP (1872–73), p. 342. Increasingly thorough enquiry and incentives to reveal increases served to confirm colonial wishes by themselves. Lord Cranbrook, for example, remarked that estimates of cotton production were often wildly wrong, partly as a result of fraud. SOSI to GOB, No. 11, 26 February 1880; NAI, HRAD, RB, March 1880, 'B' proceedings 92. Sumit Guha argues that the impact of cotton exports on the region was 'negligible' – *Agrarian Economy*, p. 54.
5. Karl Marx, *Capital, Vol. III* (Harmondsworth, 1981), pp. 215–16.
6. Havelock to GOB, No. 4799, 9 December 1873, Annual Report of Land and Sayer Revenue in Northern and Southern Divisions for 1872–73; MSA, GOB, RD, Vol. 11 of 1874, No. 62, Appendix VI.
7. GRABP (1870–71), pp. cxl–cxliii.
8. Thus the British sought to improve the output of foodgrain agriculture to support the development of commercial cotton growing in much the same way as, a century later, independent India attempted to reform the whole agricultural sector to support industrialisation by providing an increased net marketed surplus. In both cases Byres argues reasonably – but with little apparent concern for unviable smallholders' prospects in transitional labour markets – that 'the agrarian structure prevented such growth'. Terry Byres, 'Land Reform, Industrialization and the Marketed Surplus in India: An Essay on the Power of Rural Bias', in David Lehmann (ed.), *Agrarian Reform and Agrarian Reformism: Studies of Peru, Chile, China and India* (London, 1974), pp. 233–7.
9. See Banaji, 'Capitalist domination', 1377.
10. *Ahmadnagar Gazetteer*, pp. 274–5.
11. Boswell to Oliphant, No. 2132, 20 July 1875, pp. 451–3.
12. Boswell to Havelock, No. 1952, 20 July 1876, pp. 228–9.
13. See, for example, GRABP (1872–73), pp. 339–40.
14. Banaji, 'Capitalist domination', 1377–9, citing GRABP (1891–92), p. 132.
15. This was the explanation given in GOB Resolution No. 6092, RD, 27 October 1875, Land and Sayer Revenue Report on the Northern and Southern Divisions for 1873–74; MSA, GOB, RD, Vol. 12 of 1875, No. 543, pp. 450–1.
16. Colonel W. C. Anderson, Survey Commissioner, ND, to GOB, No. 409, 11 April 1881; NAI, RAD, RB, August 1881, 'B' proceedings 15.
17. See, for example, King to Robertson, No. 3140, 19–23 July 1879, pp. 23–4.
18. See correspondence in the Ahmednagar Collector's Office, Rack 10, Rumal 192, No. 15, 1878.
19. See King to Robertson, No. 4161, 20 July 1880, pp. 23–4, and also King to Robertson, No. 4584, 22–5 July 1881, p. 60.

20. King to Robertson, No. 3140, 19–23 July 1879, p. 53.
21. Boswell to Havelock, No. 1642, 20 July 1874, pp. 35–6.
22. Sen, *Poverty and Famines*, pp. 154–6.
23. Banaji, 'Capitalist Domination', 1383.
24. See, for example, King to Robertson, No. 3140, 19–23 July 1879, pp. 7–10.
25. *Bombay City Gazetteer, Vol. III* (Bombay, 1910), pp. 214–15.
26. GOB to SOSI, No. 39, 17 December 1874; NAI, RACD, LRSB, March 1875, 'A' proceedings 6–19, p. 2.
27. GRABP (1874–75), General Summary, p. xxix.
28. *Ibid.*, p. 115.
29. Woodburn to Boswell, No. 271, 5 July 1875, p. 603.
30. The inconsistency between the 1871–72 figures here and in Table 2.4 is unexplained in the source.
31. King to Robertson, No. 4584, 22–5 July 1881, p. 116.
32. King to Robertson, No. 4161, 20 July 1880, p. 26.
33. FCR (1880), 'Information and Evidence' chapter 1, question 9, answer by R. B. Mahadev Wasudev Barve.
34. Boswell to Havelock, No. 1642, 20 July 1874, p. 36.
35. John Hurd, 'Railways', in Dharma Kumar (ed.), *The Cambridge Economic History of India, Vol. 2* (Cambridge, 1983), p. 743.
36. See GRABP (1874–75), p. xliv.
37. GRABP (1875–76), pp. lxxxiii–iv.
38. FCR (1880), 'Bombay Information and Evidence', chapter 1, question 19, answer by Major-General St Clair Wilkins, Superintending Engineer, SD.
39. FCR (1880), Appendix II, p. 42.
40. Temple, 'The Famine', p. 94.
41. King to Robertson, No. 4584, 22–5 July 1881, p. 221.
42. GOI to all Provincial Governments, No. 4, 20 July 1878; NAI, GOI, RACD, LRSB, July 1878, 'B' proceedings 97.
43. Stewart to Robertson, No. 3195, 22–4 July 1878, pp. 291, 341–2.
44. King to Robertson, No. 4161, 20 July 1880, p. 114.
45. Stewart to Robertson, No. 3195, 22–4 July 1878, pp. 342–3.
46. GOB to Robertson, No. 846F, 5 May 1877; MSA, GOB, RD, Famine Branch, Vol. 24 of 1876–77, p. 529.
47. Elphinston to Robertson, No. 5730, 20 July 1882, pp. 247–9.
48. Jacomb to Robertson, No. A/4960, 19 July 1877, p. 19.
49. Stewart to Robertson, No. 3195, 22–4 July 1878, pp. 285–6.
50. King to Robertson, No. 3140, 19–23 July 1879, p. 18 and King to Robertson, No. 4161, 20 July 1880, p. 18.
51. *Ibid.*, pp. 24 and 17 respectively.
52. Woodburn to Boswell, No. 59, 5 July 1876, p. 322. Original emphasis.
53. Jacomb to Robertson, No. A/4960, 19 July 1877, Appendix B, p. 87.
54. Stewart to Robertson, No. 3195, 22–4 July 1878, p. 285 and Robertson to GOB, No. R/2777, 5 August 1878, AARA (1878), p. 542.
55. Charlesworth, 'Rich peasants', pp. 105–7.
56. King to Robertson, No. 4161, 20 July 1880, pp. 27–9.
57. Robertson to GOB, No. R/2431, 2 August 1880, pp. 326–7.
58. Elphinston to Robertson, No. 5730, 20 July 1882, p. 68.
59. *Ahmadnagar Gazetteer*, p. 343.

60. *Ibid.*, p. 70.
61. Washbrook, 'Commercialization', 141.
62. FCR (1880), Part I, pp. 66–9.
63. Hamilton to Jacomb, No. 374, 5 July 1877, p. 123 and Jacomb to Robertson, No. 4961, 19 July 1877, p. 153.
64. Hamilton to Jacomb, No. 225, 10 April 1877; MSA, GOB, RD, Famine Branch, Vol. 24 of 1876–77, p. 487, and other related correspondence, pp. 481–95.
65. Woodburn to Jacomb, No. 481, 5 July 1877, pp. 184–5 and King to himself, No. 164, 21 July 1879, AARA (1879), p. 210.
66. Woodburn to Boswell, No. 58, 5 July 1875, AARA (1875), p. 674.
67. Blathwayt to Boswell, No. 168, 7 July 1876, p. 362.
68. Boswell to Oliphant, No. 2132, 20 July 1875, p. 508.
69. *Ibid.*, p. 509.
70. W. G. Pedder, Acting Secretary to the GOB, 'Review of the Administration Reports of the Collectors, their Assistants, and District Deputies for 1874–5', 16 September 1875, AARA (1875), p. 763.
71. Robertson to Kennedy, No. 352A, 27–31 January 1877; MSA, GOB, RD, Famine Branch, Vol. 23 of 1876–77, pp. 537–9.
72. Jacomb to Robertson, No. A/4960, 19 July 1877, pp. 60–1.
73. See Candy to Stewart, No. 514, 11 July 1878, pp. 418–19.
74. GOB Resolution No. 45.C.W-109, 20 January 1877, Famine Vol. 23, p. 387.
75. Robertson to GOB, No. 3620, 3 August 1877, p. 262.
76. Hamilton to King, No. 276, 5 July 1879, p. 172.
77. FCR (1880), 'Bombay Information and Evidence,' chapter 1, question 18, answer by Colonel Jenkin Jones, Superintending Engineer, CD.
78. King to Robertson, No. 3140, 19–23 July 1879, p. 68.
79. GRABP (1872–73), p. x and p. 352.
80. See de Waal, *Famine That Kills*, p. 7; Neeladri Bhattacharya, 'Agricultural Labour and Production: Central and South-East Punjab', in K. N. Raj, N. Bhattacharya, S. Guha and S. Padhi (eds), *Essays on the Commercialisation of Indian Agriculture* (Delhi, 1985), pp. 122–3 and N. S. Jodha, 'Famine and Famine Policies: Some Empirical Evidence', *EPW*, 10 (41) (1975), 1613–14.
81. Bina Agarwal, 'Social Security and the Family: Coping with Seasonality and Calamity in Rural India', in E. Ahmad, J. Dreze, J. Hills and A. Sen (eds), *Social Security in Developing Countries* (Oxford, 1991), pp. 217–18.
82. Charlesworth, 'Rich peasants', pp. 109–13.
83. The latter phenomenon significantly reduced the proportion of Ahmednagar's marketed surplus that was actually sold in external markets. On the distinction between gross and net marketed surplus, see Byres, 'Land reform', p. 227.
84. This is consistent with Marx's assertion that the development of commercial capital (as distinct from commodity capital) 'taken by itself, is insufficient to explain the transition from one mode of production to [an]other' in part because 'when commercial capital exchanges the products of undeveloped communities, commercial profit not only appears as defrauding and cheating but to a large extent does derive precisely from this'. *Capital, Vol III*, pp. 444, 448.
85. Havelock to GOB, No. 2831, 4 August 1874, pp. 105–6.
86. Elphinston to Robertson, No. 5730, 20 July 1882, p. 90.

87. Martin Ravallion, *Markets and Famines* (Oxford, 1987) pp. 58–62.
88. GOB to SOSI, No. 39, 17 December 1874; NAI, RACD, LRSB, March 1875, 'A' proceedings 6–19, p. 2.
89. See, for example, GOB to SOSI, No. 2, 28 March 1879, NAI, HRAD, RB, October 1879, 'A' proceedings, 76–80.
90. See, for example, Jacomb to A. H. Spry, 1st. Assistant Collector and Magistrate, Ahmednagar, No. 3796, 16–18 December 1876, Famine Vol. 23, p. 656.
91. For example, successful cloth and sari manufacture for export from Sangamner *taluka* was heavily curtailed in the early 1870s. Boswell to Oliphant, No. 2132, 20 July 1875, p. 451. When the collectors were directed to list newly developed or expanding manufacture in 1881, King replied that there was none in Ahmednagar. King to Robertson, No. 4584, 22–5 July 1881, p. 59.
92. See, for example, the views of Pedder, a member of the Deccan Riots Commission, in DRCR, Appendix A, pp. 135–41.
93. Srinivasa Ambirajan, *Classical Political Economy and British Policy in India* (Cambridge, 1978), p. 100.
94. A. Mackenzie, President of the Madras Chamber of Commerce, argued that the purchase of grain could not have affected markets, provided it was sold publically and not supplied as relief in kind. He further recommended the signing of contracts with major grain merchants to guarantee supplies in the event of future famines. FCR (1880), Appendix I, pp. 203–4. For an example of the negative reaction to this policy from above, see Temple to Salisbury, 17 January 1877.
95. See FCR (1880), Part III, p. 125.
96. Compilations of Native Newspaper Reports, Bombay Presidency (NNR), *Nagar Samachar*, 14 October 1876.
97. Jacomb to Robertson, No. A/4960, 19 July 1877, p. 20.
98. Sen, *Poverty and Famines*, pp. 3–4.
99. In November 1876 Woodburn reported that 'feuds among villagers' had closed the market at Mirajgaon, which was the sole source of supply for two *talukas*, and that the collector 'could not send too much grain to Shrigonda and Karjat'. When imports started, however, the bazaars relaxed and Mirajgaon stocks came onto the market. Cited in Jacomb to Robertson, No. 1550, 6 March 1877, Famine Vol. 63, p. 118.
100. Dissenting Famine Commissioners Sullivan and Caird, pointing out that 'trade as a rule acts cautiously, and is not influenced by sentiment', argued that price rises regarded by the government as an early warning of famine were too uncertain for external traders to gamble on. They would only regard exporting to scarcity districts as viable once 'the uncertainty has become reality' as a result of starvation deaths. FCR (1880), part I, p. 68.
101. See Washbrook, 'Commercialization', 162–3.
102. Jacomb to Robertson, No. A/4960, 19 July 1877, p. 170.
103. Major H. Daniell, District Superintendent of Police, Ahmednagar, claimed he had personally tried to persuade small shops to re-open, including those without stocks, so that people would not suspect them of hoarding. Interestingly, this was in response to a damning rumour that he had himself closed shops which were charging too much. Daniell to GOB, No. 36, 23 January 1877; MSA, GOB, RD, Famine Branch, Vol. 81 of 1876–77, pp. 173–5.

104. See FCR (1880), part III, p. 78.
105. Cited in Ambirajan, 'Political Economy', 27.
106. Cited in Bhatia, *Famines in India*, p. 106.
107. *Ibid.*, p. 31.
108. Report on the History of the Famine in His Highness the Nizam's Dominions (Bombay, 1879), p. 64.
109. Stewart to Robertson, No. 3195, 22–4 July 1878, p. 288.
110. GRABP (1877–78), pp. v–vi.
111. Jacomb to Robertson, No. A/4960, 19 July 1877, p. 21.
112. GRABP (1876–77), p. lviii.
113. Temple, 'Regulation of Grain Traffic on the Great Indian Peninsula Railway for the Distressed Districts', *Minutes 1877–80, Vol. I.*
114. GRABP (1877–78), p. vii.
115. Jacomb to Robertson, No. A/4960, 19 July 1877, p. 24.
116. Jenkin Jones to GOB, No. 38D, 24 November 1876, Famine Vol. 23, pp. 148–9.
117. GOB to Jenkin Jones, No. 717, 1 December 1876, Famine Vol. 23, p. 151.
118. Jenkin Jones to GOB, No. 78D, 8 December 1876, Famine Vol. 23, pp. 156–60.
119. Colonel Iver, Acting Superintending Engineer, CD, to GOB, No. 650, 21 February 1877, Famine Vol. 62, p. 544.
120. GOB Resolutions No. 61 E-206, 29 January 1877, and No. 381F, 23 February 1877, Famine Vol. 62, pp. 355, 549 respectively.
121. Jacomb to Robertson, No. 1550, 6 March 1877, Famine Vol. 63, pp. 114–5.
122. Candy to Jacomb, No. 299, 22 May 1877, Famine Vol. 24, pp. 635–6.
123. Robertson to GOB, No. 2475, 25 May 1877, Famine Vol. 24, p. 639.
124. Candy to Jacomb, No. 471, 13 July 1877, Famine Vol. 63, pp. 551–2.
125. Hamilton and Fforde, jointly to Jacomb, No. 76, 1 February 1877, Famine Vol. 62, p. 409.
126. Jacomb to Robertson, No. 5563, 15 August 1877, p. 111; Robertson to GOB, No. 3880, 20 August 1877, p. 113; and GOB Resolution No. 394P, 25 August 1877; MSA, GOB, RD, Famine Branch, Vol. 25 of 1876–77, p. 117.
127. See, for example, Woodburn's suggestion that grain dealers made 'fortunes'. Woodburn to Stewart, No. 291, 5 July 1878, p. 501.
128. Bhatia, *Famines in India*, p. 107.
129. Robertson to GOB, No. R/2401, 19 August 1879, p. 259.
130. Hamilton to King, No. 592, 9 July 1881, p. 307.
131. McAlpin, *Subject to Famine*, looks at western India between the former dates.
132. Washbrook points out that not only were fewer foodgrains grown in more commercially advanced areas, local magnates sold off food stocks on good terms prior to the famine. Washbrook, 'Commercialization', 141.

3 Rural Moneylending, Credit Legislation and Peasant Protest

1. This was noted by Pedder, 'Review of reports', 16 September 1875, p. 782.
2. DRCR, p. 23.
3. Neil Charlesworth, 'The Myth of the Deccan Riots', in Hardiman, *Peasant Resistance*, pp. 219–21.
4. Havelock to GOB, No. 2831, 4 August 1874, pp. 104–5.

5. *Ahmadnagar Gazetteer*, p. 242.
6. DRCR, Appendix A, pp. 8–9, citing 'Papers Relating to the Revision of Assessment in Six Talukas of the Ahmednagar Collectorate', GOB Records No. CXXIII New Series, pp. 1851–3.
7. Boswell to Havelock, No. 2132, 20 July 1875, p. 487.
8. DRCR, pp. 58–9.
9. *Ibid.*, p. 22.
10. Woodburn to Boswell, despatching Loch's report, No. 271, 5 July 1875, p. 642.
11. FCR (1880), part II, p. 131.
12. *Ibid.*, p. 132.
13. DRCR, p. 22.
14. Memorandum by E. C. Buck, Secretary, GOI, RAD, 25 May 1882; NAI, GOI, RAD, RB, June 1884, 'A' proceedings 13–26, Keep-With Section, p. 4.
15. See, for example, FCR (1880), 'Information and Evidence', chapter 1, question 9, answer by Barve.
16. Charlesworth, 'Rich Peasants', pp. 97–100.
17. See, for example, Ravinder Kumar, 'Deccan Riots', p. 155.
18. Sumit Guha, *Agrarian Economy*, p. 75.
19. *Ahmadnagar Gazetteer*, p. 315.
20. Sumit Guha, *Agrarian Economy*, pp. 70–8.
21. DRCR, p. 23.
22. *Ibid.*, p. 62.
23. *Ahmadnagar Gazetteer*, pp. 299–301.
24. Charlesworth, 'Rich Peasants', pp. 99–101.
25. Sumit Guha, *Agrarian Economy*, p. 73.
26. Cited in Charlesworth, 'The Myth', p. 226.
27. Cited in DRCR, pp. 45–6.
28. Pedder, 'Review of Reports', 16 September 1875, p. 762.
29. Catanach, *Rural Credit*, p. 16.
30. DRCR, pp. 25–6.
31. Cited in Kenneth Ballhatchet, *Social Policy and Social Change in Western India, 1817–1830* (London, 1957), p. 153.
32. Hardiman, *Feeding the Baniya*, pp. 51–2.
33. Andre Wink, *Land and Sovereignty in India: Agrarian Society and Politics under the Eighteenth-century Maratha Svarajya* (Cambridge, 1986), pp. 334–5 and Frank Perlin, 'Of White Whale and Countrymen in the Eighteenth Century Maratha Deccan: Extended Class Relations, Rights and the Problem of Rural Autonomy under the Old Regime', *Journal of Peasant Studies*, 5 (2) (1978), 190.
34. DRCR, p. 28.
35. Cited *ibid.*, pp. 45–6.
36. Quoted in Boswell to Havelock, No. 1645, 20 July 1874, p. 121.
37. DRCR, p. 98.
38. Hardiman, *Feeding the Baniya*, pp. 187–9.
39. W. F. Sinclair, Assistant Collector, Ahmednagar, to Boswell, No. 342, 6 September 1875, reprinted in DRCR, Appendix A, p. 299.
40. Hardiman, *Feeding the Baniya*, p. 53.
41. Cited in K. G. Sivaswamy, *Legislative Protection and Relief of Agriculturist Debtors in India* (Poona, 1939), p. 33.
42. DRCR, p. 102.

43. C. E. Fraser Tytler, Collector of Ahmednagar, 1858, cited in DRCR, Appendix A, p. 17.
44. DRCR, pp. 63–4.
45. Marx, *Capital, Vol. III*, p. 321.
46. Banaji, 'Capitalist Domination', 1376.
47. FCR (1880), Appendix I, p. 180.
48. DRCR, pp. 24, 33.
49. Cited in Ravinder Kumar, 'Deccan Riots', p. 157.
50. GOB Resolution No. 4396, 31 July 1873; MSA, GOB, RD, Vol. 26 of 1873, No. 1091.
51. Charlesworth, 'The Myth', pp. 209–14; Morris D. Morris, 'Economic Change and Agriculture in Nineteenth Century India', *IESHR*, 3 (2) (1966), 191; Dharma Kumar, *Land and Caste in South India: Agricultural Labour in the Madras Presidency in the Nineteenth Century* (Cambridge, 1965), p. 179; Bernard Cohn, 'Recruitment of Elites in India under British rule', Conference Paper, 1968. All cited in Catanach, *Rural Credit*, pp. 12–13.
52. DRCR, p. 53.
53. *Ibid.*, pp. 59–60. Loch also researched land ownership and found 1182 holdings in *Marwaris'* names in Parner, Sangamner and Akola *talukas*. Woodburn to Boswell, No. 271, 5 July 1875, p. 605.
54. For similar evidence, see Banaji, 'Capitalist Domination', 1385–6.
55. Charlesworth, 'Rich Peasants', pp. 110–11.
56. DRCR, Appendix A, Evidence of Boswell, p. 254.
57. Jacomb to Robertson, No. A/4960, 19 July 1877, p. 37.
58. Havelock to GOB, No. 2831, 4 August 1874, p. 107.
59. DRCR, p. 57.
60. DRCR, pp. 83–5.
61. Blathwayt to Boswell, No. 168, 7 July 1876, p. 350. He also reported (p. 353) the need to sack a village accountant who was implicated directly in rioting against *sowcars*.
62. FCR (1880), 'Information and Evidence', chapter 1, question 10, answer by Robertson concerning Ahmednagar.
63. Boswell to Havelock, No. 2132, 20 July 1875, p. 487.
64. *Ibid.*, pp. 463–4. Conversely, Hardiman suggests that *Marwaris* also colluded at auctions to ensure that their kinsmen acquired property – which had become available because of their own refusal to pay their clients' taxes – cheaply. *Feeding the Baniya*, pp. 190–1. Where the property was land, this could have been an alternative way of evicting *ryots* without recourse to interfering courts.
65. Stokes, *Peasant and Raj*, p. 12.
66. DRCR, p. 93.
67. This issue arose in Kaira district. See GOI to GOB, No. 150R, 19 April 1882; NAI, GOI, RACD, RB, April 1882, 'A' proceedings 13–14.
68. Stewart to Robertson, No. 3195, 22–4 July 1878, pp. 296–7.
69. Hardiman, *Feeding the Baniya*, pp. 214–19.
70. Catanach, 'Agrarian Disturbances', pp. 189–91.
71. Boswell to Havelock, No. 2132, 20 July 1875, p. 498.
72. GOB to GOI, No. 2202, 6 April 1877, cited in FCR 1880, 'Information and Evidence', chapter 1, question 10, answer by Robertson.

73. DRCR, Vol. II, p. 4.
74. DRCR, Appendix B, pp. 1–45.
75. DRCR, pp. 83–5.
76. FCR (1880), part II, pp. 130–6.
77. Charlesworth, 'The Myth', p. 223, suggests that there was a dominant view in the GOB, of which the Deccan Riots Commission Chairman, J. B. Richey, was part, that it should be allowed to make active interventions to reduce indebtedness, while another member, Auckland Colvin of Bengal, had risen through the United Provinces and was thus keen to criticise the *ryotwari* revenue system. The hand-picking of the Famine Commission to take a more consistent state line is discussed in Brennan, 'Famine Code', pp. 97–101.
78. DRCR, p. 108.
79. *Ibid.*, p. 4.
80. *Ibid.*, p. 5.
81. Stewart to Robertson, No. 3195, 22–4 July 1878, p. 333.
82. This was between April 1871 and October 1874. DRCR, p. 9.
83. *Ahmadnagar Gazetteer*, p. 420.
84. Hardiman, *Feeding the Baniya*, p. 271.
85. *Ahmadnagar Gazetteer*, p. 420. Though *Kunbi ryots* were indeed rarely violent, their general support for *dacoits* – and even that of some village law enforcement officers – was noted with concern in King to Robertson, No. 4161, 20 July 1880, p. 88.
86. Hardiman, *Feeding the Baniya*, pp. 254–5, in which he cites, *inter alia*, James C. Scott, *The Moral Economy of the Peasant: Rebellion and Subsistence in Southeast Asia* (New Haven, 1976); E. P. Thompson, 'The Moral Economy of the English Crowd in the Eighteenth Century', *Past and Present*, 50 (1971), 76–136 and David Arnold, 'Looting, Grain Riots and Government Policy in South India 1918', *Past and Present*, 84 (1979), 111–45.
87. Ravinder Kumar, 'Deccan Riots', pp. 168–9.
88. Alexander de Waal, *Famine Crimes: Politics and the Disaster Relief Industry in Africa* (Oxford, 1997), pp. 12–16.
89. Scott, *Moral Economy of the Peasant*; Thompson, 'English Crowd'.
90. DRCR, Appendix A, evidence, respectively, of Boswell, p. 255, and Norman, p. 246.
91. Charlesworth, *Peasants and Imperial Rule*, p. 115.
92. Boswell to Havelock, No. 2132, 20 July 1875, p. 498.
93. *Ibid.*, pp. 440–6.
94. *Ibid.*, p. 501.
95. Boswell to Havelock, No. 1952, 20 July 1876, p. 231.
96. DRCR, p. 116.
97. Memorandum by Richey, 15 September 1875, Demi-official correspondence relating to the Deccan Riots; MSA, GOB, RD, Vol. 118 of 1875, No. 1867, pp. 311–444.
98. Memorandum by Hon. A. Rogers, 19 September 1875; *ibid.*
99. Memorandum by Governor P. E. Wodehouse, 17 September 1875; *ibid.*
100. King to Robertson, No. 4161, 20 July 1880, pp. 133–5.
101. King to Robertson, No. 4584, 22–5 July 1881, pp. 245–7.

102. Sivaswamy, *Legislative Protection*, pp. 38–51.
103. Douglas Haynes, 'Market Formation in Khandesh, c.1820–1920', *IESHR*, 36 (3) (1999), 298.
104. DRCR, pp. 114–15.
105. King to Robertson, No. 4161, 20 July 1880, pp. 29–30, 44, 55.
106. *Ibid.*, p. 29.
107. GOB Resolution, No. 4791, 10 September 1880, AARA (1880), p. 368.
108. Robertson to GOB, No. R/2431, 2 August 1880, p. 327.
109. Hamilton to King, No. 592, 9 July 1881, p. 308.
110. Elphinston to Robertson, No. 5730, 20 July 1882, pp. 106–7.
111. *Ibid.*, pp. 85–7.
112. King to Robertson, No. 4584, 22–5 July 1881, p. 78.
113. *Ibid.* Original emphasis.
114. Keen, *Benefits of Famine*, p. 232.
115. DRCR, p. 49.
116. *Ibid.*, p. 47.
117. King to Robertson, No. 3140, 19–23 July 1879, p. 42.
118. Pedder, 'Review of Reports', 16 September 1875, p. 782.
119. King to Robertson, No. 4161, 20 July 1880, p. 55.
120. DRCR, Appendix A, Evidence of Boswell, p. 260.
121. FCR (1880), Appendix I, p. 180.
122. Boswell to Havelock, No. 2132, 20 July 1875, pp. 469–70.
123. Banaji, 'Capitalist Domination', 1387.
124. David Washbrook, 'Progress and Problems: South Asian Economic and Social History c.1720–1860' *MAS*, 22 (1) (1988), 88–91.
125. GOB Resolution, No. 6711, 13 December 1873; MSA, GOB, RD, Vol. 26 of 1873, No. 1091.
126. Catanach, 'Agrarian Disturbances', p. 191.
127. Settlement loans were given to both *Bhils* and *Kunbis* who would move to *Bhil* regions. See, for example, GOB Resolution No. 4160, regarding Khandesh, 28 August 1871; NAI, GOI, RACD, Tuccavee Branch, September 1871, 'A' proceedings 11–12. On interest free *takavi* for government servants, see the summary of *takavi* rules in NAI, GOI, RACD, General Branch, August 1877, 'A' proceedings 1.
128. Whitley Stokes, GOI, Legislative Department, to GOI, RACD, No. 59, 1 February 1873; NAI, GOI, RACD, Tuccavee Branch, October 1875, 'A' proceedings 1–22.
129. FCR (1880), Appendix II, Evidence of Temple, p. 32.
130. GOB to GOI, No. 1425, 26 March 1872; NAI, GOI, RACD, Tuccavee Branch, August 1872, 'A' proceedings 1–25 and GOI to GOB, No. 72, 9 November 1872; NAI, GOI, RACD, Tuccavee Branch, November 1872, 'B' proceedings 1–2.
131. E. W. Ravenscroft, Acting Chief Secretary, GOB, to GOI, No. 6408, 21 December 1871; NAI, GOI, RACD, Tuccavee Branch, November 1872, 'B' proceedings 4.
132. FCR (1880), part II, p. 144.
133. GOB to GOI, No. 6683, 10 November 1881; NAI, GOI, RAD, RB, October 1882, 'A' proceedings 28–50.
134. Elphinston to Robertson, No. 5732, 20 July 1882, AARA 1882, p. 475.
135. FCR (1880), 'Information and Evidence', chapter 1, question 11, answer by Colonel Anderson.

136. GOB to GOI, No. 3775, 5 July 1875; MSA, GOB, RD Vol. 149 of 1875, No. 81. See also correspondence in RD Vol. 136 of 1874, No. 81.
137. H. Erskine, Collector of Nasik, to Rogers, Revenue Commissioner, ND, No. 1150, 6 April 1874; MSA, GOB, RD Vol. 136 of 1874, No. 81.
138. Ibid.
139. Rogers to GOB, *ibid*.
140. King to Robertson, No. 3140, 19–23 July 1879, pp. 20–1.
141. Stokes to GOI, RACD, No. 59, 1 February 1873.
142. Bombay Rule XXIII under the Land Improvement Act, 1871; NAI, GOI, RACD, Tuccavee Branch, December 1873, appendix B to Proceedings Volume for July 1871 to December 1876.
143. GOB Resolution, No. 756, 5 February 1881, discussed in King to Robertson, No. 4584, 22–5 July 1881, p. 57, and GOB Resolution, No. 7244, 30 November 1881, AARA (1881), pp. 501–2.
144. FCR (1880), part II, p. 144.
145. Bombay Rule IX under the Land Improvement Act, 1871.
146. Memoranda by T. W. Holderness, Officiating Secretary, RAD, 20 April 1883, and Sir Steuart Bayley, RAD Member of the Governor-General's Council, 18 May 1883; NAI, GOI, RAD, RB, March 1884, 'A' proceedings 1, Keep-With Section.
147. Accountant General of Bombay to GOB, RD, No. 7057, 18 August 1875; MSA, GOB, RD Vol. 149 of 1875, No. 81.
148. Jacomb to Robertson, No. A/4960, 19 July 1877, p. 22.
149. Robertson to GOB, No. 3619, 3 August 1877, AARA (1877), p. 77.
150. Memorandum by C. P. Ilbert, Legislative Member, 16 September 1882; NAI, GOI, RAD, RB, October 1882, 'A' proceedings 28–50, Keep-With Section.
151. McAlpin, *Subject to Famine*, pp. 179–84.
152. GOB to GOI, No. 6683, 10 November 1881.
153. GOI to GOB, No. 22, 7 October 1871, and Ashburner, Collector of Khandesh, to Havelock, No. 2527, 22 August 1871; NAI, GOI, RACD, Tuccavee Branch, October 1871, 'A' proceedings 1–2.
154. Pedder to GOI, No. 3775, 5 July 1875.
155. FCR (1880), part II, p. 144.
156. Memorandum by Holderness, 13 February 1882; NAI, GOI, RAD, RB, October 1882, 'A' proceedings 28–50, Keep-With Section.
157. GOB Resolution, No. 6028, 23 October 1876. A breakdown of expenditure of this sum, totalling 13,124 rupees by March 1877, is given in Jacomb to Robertson, No. A/4960, 19 July 1877, p. 23.
158. Hamilton to Jacomb, No. 374, 5 July 1877, p. 119.
159. Woodburn to Jacomb, No. 481, 5 July 1877, p. 160.
160. *Ibid.*, p. 161.
161. Bhatia, *Famines in India*, p. 120.
162. FCR (1880), Appendix II, pp. 55–6.
163. GOB to GOI, No. 6683, 10 November 1881.
164. GOB, PWD Resolution, No. 99P, 21 June 1877, cited in Stewart to Robertson, No. 3195, 22–24 July 1878, p. 289.
165. Candy to King, No. 362, 7–11 July 1879, AARA (1879), p. 133.
166. Elphinston to Robertson, No. 5730, 20 July 1882, pp. 75–9.
167. Robertson to GOB, No. R/3701, 9 August 1882, p. 309.

168. King to Robertson, No. 3140, 19–23 July 1879, p. 22.
169. King to Robertson, No. 3141, 19–23 July 1879, p. 141.
170. King to Robertson, No. 3142, 19–23 July 1879, AARA (1879), p. 187.
171. Candy to King, No. 561, 12 July 1880, p. 191.
172. FCR (1880), part II, p. 144.
173. Memorandum by Holderness, 16 March 1882; NAI, GOI, RAD, RB, October 1882, 'A' proceedings 28–50, Keep-With Section, p. 10.
174. Act XIX of 1883 was passed on 12 October 1883 and circulated to local governments with GOI No. 83R, 30 October 1883; NAI, GOI, RAD, RB, March 1884, 'A' proceedings 1.
175. SOSI to GOI, No. 5, 31 January 1884; NAI, GOI, RAD, RB, April 1884, 'A' proceedings 6–8.
176. Memorandum by Major E. Baring, Finance Member, GOI, 17 August 1882; NAI, GOI, RAD, RB, October 1882, 'A' proceedings 28–50, Keep-With Section, p. 32.
177. FCR (1880), part II, p. 145.
178. Woodburn to Stewart, No. 291, 5 July 1878, p. 502.
179. McAlpin, *Subject to Famine*, p. 181.
180. DRCR, Appendix A, pp. 76–8.
181. GOB Resolution, No. 870, 3 March 1860, cited in *ibid.*, pp. 108–9.
182. FCR (1880), Appendix I, pp. 191–7.
183. NAI, GOI, RAD, RB, January 1883, 'A' proceedings 13–16.
184. Memorandum by W. Lee-Warner, Bombay Civil Service, 29 April 1882; NAI, GOI, RAD, RB June 1884, 'A' proceedings 13–26, Keep-With Section p. 5.
185. William Wedderburn, Judge, Ahmednagar, to Under-SOSI, 13 December 1883; *ibid.*
186. GOB Resolution, No. 7272, 18 October 1882, AARA (1882), p. 630.
187. Memorandum by Baring, 17 August 1882, p. 33.
188. GOB to GOI, No. 2687, 3 April 1883; NAI, GOI, RAD, RB, June 1884, 'A' proceedings 13–26.
189. GOB to GOI, No. 5162, 26 June 1884; NAI, GOI, RAD, RB, July 1884, 'A' proceedings 19–20.
190. Stewart, Survey and Settlement Commissioner, to GOB, No. 2481, 13 December 1882; NAI, GOI, RAD, RB, June 1884, 'A' proceedings 13–26.
191. Memorandum by T. C. Hope, RAD, 20 September 1883; NAI, GOI, RAD, RB, June 1884, 'A' proceedings 13–26, Keep-With Section.
192. Memorandum by Buck on the first meeting of the GOI's Agricultural Bank Committee, 17 October 1882; NAI, GOI, RAD, RB, January 1883, 'A' proceedings 13–16.
193. *Ibid.*
194. Memorandum by Baring, 17 August 1882, pp. 35–6.
195. GOI to SOSI, No. 7, 31 May 1884; NAI, GOI, RAD, RB, June 1884, 'A' proceedings 13–26.
196. SOSI to GOI, No. 95, 23 October 1884; NAI, GOI, RAD, RB, December 1884, 'A' proceedings 68–70.
197. Krishna Chandra to GOI, 15 January 1883; NAI, GOI, RAD, RB, February 1883, 'B' proceedings 53.
198. Catanach, *Rural Credit*, pp. 25–6.
199. Rosalind O'Hanlon, *Caste, Conflict and Ideology: Mahatma Jotirao Phule and Low Caste Protest in Nineteenth-Century Western India* (Cambridge, 1985), p. 260.

4 Land Revenue Rigidity, Revisions and Non-remission

1. B. H. Baden-Powell, *The Land Systems of British India, Volume 1* (Oxford, 1892), p. 26.
2. FCR (1880), part II, p. 89.
3. *Ibid.*, p. 90.
4. FCR (1880), Appendix I, p. 158.
5. GRABP (1876), p. lxxxvii.
6. GRABP (1873), p. 444. Figures converted from rupees at the universal exchange rate of ten to the pound.
7. GRABP (1876), p. lxxxvii.
8. Minute by Lord Salisbury, SOSI, 1875, cited by Bhatia, *Famines in India*, p. 23.
9. Bhatia, *Famines in India*, p. 309 and Stokes, *English Utilitarians*, p. 318.
10. GOI Resolution, No. 2, 18 March 1909, quoted in Bhatia, *Famines in India*, p. 309.
11. GRABP (1871), pp. 89–91.
12. GRABP (1872), pp. 139–42.
13. GRABP (1873), p. 491.
14. GOB to GOI, No. 3362, 7 June 1876; MSA, GOB, RD, Vol. 49, p. 224.
15. GRABP (1873), p. 492.
16. Speech by Sir John Strachey, *Indian Famine Speeches* (Delhi, 1879), p. 19.
17. Jacomb to Robertson, No. A/4960, 19 July 1877, p. 72.
18. GRABP (1872), pp. 145–7.
19. GRABP (1873), pp. xxv–xxvi.
20. FCR (1880), part II, p. 92. There were two shillings to the rupee.
21. FCR (1880), part II, p. 112, cited by Romesh Dutt, *Open Letters to Lord Curzon on Famines and Land Assessments in India* (London, 1900), p. 97.
22. DRCR, Vol. II, Minute by A. Lyon, Bombay Civil Service, p. 17.
23. GRABP (1874), p. lv.
24. House of Commons answer by Mr Cross, Under-Secretary of State for India, 16 August 1883; NAI, RAD, RB, September 1883, 'B' proceedings 34.
25. SOSI to GOB, No. 34, 4 December 1879; NAI, HRAD, RB, January 1880, 'A' proceedings 57–8.
26. Handwritten memorandum by Bayley, 23 August 1882; NAI, RAD, RB, July 1883, 'A' proceedings 47–9, Keep-With section, pp. 7–8.
27. Stokes, *English Utilitarians*, p. 82.
28. *Ibid.*, pp. 42 and 75–8.
29. Ravinder Kumar, *Western India*, pp. 318–26.
30. FCR (1880), part II, p. 184.
31. GRABP (1875), p. xiv.
32. Quoted in DRCR, pp. 18–19. See also Sumit Guha, *Agrarian Economy*, pp. 28–31, citing R. E. Frykenberg, *Guntur District, 1788–1848: A History of Local Influence and Central Authority in South India* (Oxford, 1965).
33. GRABP (1875), p. xiv.
34. Sumit Guha, *Agrarian Economy*, pp. 195–200.
35. McAlpin, *Subject to Famine*, pp. 109–40.
36. *Ibid.*, pp. 198–202 and 211–12, citing Romesh Dutt, *The Economic History of India*, Vol. I (London, 1902), including the introduction by D. R. Gadgil. Bhatia rather makes the case that lower land revenue in the twentieth

century helped to end famine than that it contributed to famine earlier. *Famines in India*, pp. 295-8.
37. McAlpin, *Subject to Famine*, pp. 131-40.
38. Charlesworth, 'Rich Peasants', pp. 97-113.
39. Rabitoy, 'System v Expediency', 529-46 and Washbrook, 'Law, State', 661-6.
40. Rabitoy, 'System v Expediency', 529.
41. Charlesworth, 'Rich Peasants', pp. 112-13.
42. Baden-Powell, *Land Systems*, p. 319.
43. GRABP (1873), p. 42.
44. Dutt, *Open Letters*, pp. 18-19.
45. Bayly, *Indian Society*, pp. 173-4.
46. See Western India Association to GOB, undated; MSA, GOB, RD, Vol. 95 of 1874, No. 588.
47. Opinion of Andrew Scoble, Advocate General of Bombay, *ibid*.
48. Memorandum by Arthur Hobhouse, 4 June 1873; NAI, RACD, LRSB, October 1873, 'A' proceedings 14-25, Keep-With section.
49. Opinion of Wodehouse; MSA, GOB, RD, Vol. 95 of 1874, No. 588.
50. GOB Resolution, No. 7272, 18 October 1882, p. 633.
51. NNR, *Jame Jamsed*, 30 April 1877.
52. McAlpin, *Subject to Famine*, p. 123.
53. DRCR, p. 47. Original emphasis.
54. For a particularly strong argument that the weight and rigidity of the revenue demand was responsible for the penetration of *Marwari* moneylenders into the Deccan, see Brahma Nand, 'The Deccan Peasant Uprising, 1875', unpublished M.Phil. thesis, Jawaharlal Nehru University, New Delhi (1980), pp. 196-202.
55. See, for example, Pedder, 'Review of Reports', 16 September 1875, p. 763.
56. SOSI to GOB, No. 34, 4 December 1879.
57. Baden-Powell, *Land Systems*, p. 369.
58. *Ibid*., p. 370.
59. F. G. H. Anderson, *Facts and Fallacies About the Bombay Land Revenue System* (Poona, 1929), pp. 132-4.
60. DRCR, p. 108.
61. GOB Resolution, No. 2619, 26 March 1884; NAI, RAD, RB, May 1884, 'A' proceedings 37-8.
62. DRCR, Appended Evidence of Havelock, p. 9.
63. Memorandum by W. W. Hunter, 20 July 1880; NAI, HRAD, RB, October 1880, 'A' proceedings 13-27, Keep-With section.
64. Romesh Dutt 'The Indian land question', in Romesh Dutt (ed.), *Land Problems in India* (Madras, 1903), p. 18. Original emphasis.
65. GOB Resolution, No. 3613, 23 August 1865; Ahmednagar Collector's Office, Rack 10, Rumal 164, No. 2.
66. D'Oyly to the Revenue Commissioner, SD, No. 1655, 19 September 1865, *ibid*.
67. Elphinston to Robertson, No. 3596, 3 June 1884, Ahmednagar Collector's Office, Rack 10, Rumal 164, No. 1.
68. Robertson to Elphinston, No. R 5268, 27 October 1884; Ahmednagar Collector's Office, Rack 10, Rumal 184, No. 71.
69. FCR (1880), part II, p. 127.
70. GOB to GOI, No. 2673, 3 April 1883; NAI, RAD, RB, July 1884, 'A' proceedings 21-32.

71. GOB Resolution, No. 1372, 17 March 1874; MSA, GOB, RD Vol. 58 of 1874, No. 757.
72. GOB Resolution, No. 2183, 29 April 1874; *ibid.*
73. Havelock to GOB, No. 1466, 20 April 1874; NAI, RACD, LRSB, March 1875, 'A' proceedings 6–19, p. 16.
74. GOB to SOSI, No. 9, 26 May 1881; NAI, HRAD, RB, June 1881, 'B' proceedings 7.
75. *Ibid.*
76. S. H. Chiplonkar, editor, *Poona Sarvajanik Sabha Quarterly Journal*, to Wedderburn, 31 August 1881; NAI, RAD, RB, October 1882, 'B' proceedings 56.
77. House of Commons question by Mr Woodall to Cross, 16 August 1883; NAI, RAD, RB, September 1883, 'B' proceedings 34.
78. Robertson to GOB, No. R/817A, 25 February 1882; MSA, Annual Land and Sayer Report on the Central Division, 1880–81, Publication No. 14590.
79. SOSI to GOB, No. 2, 8 February 1883; NAI, RAD, RB, April 1883, 'A' proceedings 21.
80. Memorandum by Buck, 20 August 1882; NAI, RAD, RB, July 1883, 'A' proceedings 47–9, Keep-With section.
81. Memorandum by Buck, 26 February 1882; NAI, RAD, RB, October 1882, 'A' proceedings 28–50, Keep-With section, p. 4.
82. See, for example, Temple, 'The Condition of the Peasantry in the Central Deccan (Districts of Poona, Ahmednagar, Sholapur and Satara)', 29 October 1878, *Minutes 1877–80, Vol. II*, p. 160.
83. Baden-Powell, *Land Systems*, pp. 320, 338.
84. T. H. Stewart, Survey and Settlement Commissioner, to GOB, No. 1360, 27 June 1883; NAI, RAD, RB, November 1883, 'A' proceedings 17–21.
85. Stokes, *English Utilitarians*, pp. 107–8.
86. Wingate and Goldsmid, 'Joint Report on the Bombay Survey and Settlement', 17 October 1840, cited *ibid.*
87. FCR (1880), Appendix II, p. 36.
88. See King to Robertson, No. 4161, 20 July 1880, p. 35.
89. GOB Resolution, No. 4359, 24 November 1866; MSA, RD, Vol. 58 of 1874, No. 757.
90. For example, GOB, RD Resolution No. 2619, 26 March 1884, justified ongoing revisions 'partly with regard to the increase in the value of the land from a rise in agricultural profits'.
91. See correspondence in NAI, GOI, RACD, LRSB, July 1878, 'B' proceedings 34–42, and GOI, RAD, RB, February 1882, 'A' proceedings 5–6.
92. Bombay Irrigation Bill, 1879, Section 45, quoted in memorandum from GOI, RACD to GOI, PWD; NAI, RACD, LRSB, April 1879, 'A' proceedings 6–7.
93. GOI to GOB, 24 March 1879; NAI, RACD, LRSB, June 1879, 'A' proceedings 18.
94. SOSI to GOI, No. 2, 5 January 1880; NAI, HRAD, RB, February 1880, 'A' proceedings 95–6, para. 4.
95. GOB Resolution, No. 2619, 26 March 1884, paras 19–20.
96. House of Commons answer by Cross, 16 August 1883.
97. GOI to GOB, No. 222, 26 February 1879; NAI, HRAD, RB, October 1880, 'A' proceedings 13–27, Keep-With section.
98. Baden-Powell, *Land Systems*, p. 344.

99. SOSI to GOB, No. 36, 16 September 1880; NAI, HRAD, RB, November 1880, 'A' proceedings 33.
100. William Wedderburn, *A Permanent Settlement for the Deccan* (Bombay, 1880) and 'Mr. Wedderburn and His Critics on a Permanent Settlement for the Deccan', *Poona Sarvajanik Sabha Quarterly Journal*, 3 (3) (1882), 35.
101. See, for example, FCR (1880), Appendix II, evidence of Joshi, p. 83.
102. GOB to SOSI, No. 39, 17 December 1874, para. 4.
103. Robertson to GOB, No. R/1957, 28 May 1881; NAI, RAD, RB, September 1881, 'B' proceedings 19.
104. GOI to GOB, No. 222, 26 February 1879, para. 16.
105. Memorandum by Buck, 19 July 1882; NAI, RAD, RB, July 1883, 'A' proceedings 47–9, Keep-With section.
106. Havelock to GOB, No. 713, 28 February 1874; NAI, RACD, LRSB, July 1874, 'B' proceedings 120, p. 18.
107. Baden-Powell, *Land Systems*, p. 370.
108. 'Mr. Wedderburn and His Critics', 35.
109. Stewart quoted in GOB to GOI, No. 6340, 27 August 1883; NAI, RAD, RB, November 1883, 'A' proceedings 17–21.
110. GOB to SOSI, No. 9, 4 December 1884; NAI, RAD, RB, December 1884, 'B' proceedings 84.
111. 'Mr. Wedderburn and His Critics', 35.
112. Dutt, *Open Letters*, pp. 7–9.
113. For example, confidential memoranda by Buck, 25 and 30 May 1882, in which he condemned many of the principles of the Bombay settlement system, implying it was not 'based on agricultural facts'. NAI, RAD, RB, June 1884, 'A' proceedings 13–26, Keep-With section.
114. GOB to SOSI, No. 9, 4 December 1884.
115. GOB to SOSI, No. 39, 17 December 1874, p. 2.
116. Quoted in DRCR, p. 19.
117. See for example Colonel J. T. Francis, Survey and Settlement Commissioner, ND, to GOB, No. 1485, 19 September 1874; NAI, RACD, LRSB, March 1875, 'A' proceedings 6–19, p. 7.
118. Francis to Havelock, No. 444, 20 March 1874; *ibid.*, p. 23.
119. Francis to GOB, No. 1485, 19 September 1874; *ibid.*, p. 10.
120. Havelock to GOB, No. 1466, 20 April 1874, p. 16.
121. GOB to GOI, No. 6674, 15 December 1879; NAI, HRAD, RB, June 1880, 'A' proceedings 76.
122. W. C. Anderson to GOB, No. 1072, 9 October 1879; *ibid.*
123. See SOSI to GOB, No. 14, 31 March 1881; NAI, HRAD, RB, June 1881, 'A' proceedings 36.
124. *Ibid.*
125. Stewart to GOB, No. 1714, 3 September 1882; NAI, RAD, RB, November 1883, 'A' proceedings 17–21.
126. Boswell to Havelock, No. 1645, 20 July 1874, pp. 136–7.
127. Boswell to Oliphant, No. 2132, 20 July 1875, pp. 489–90.
128. *Ibid.*, p. 489.
129. *Ibid.*, p. 491.
130. *Ibid.*, p. 492.
131. DRCR, Appendix A, pp. 221–8.

132. DRCR, Appended evidence of unnamed Sub-Judge of Ahmednagar, p. 1.
133. Boswell to Havelock, No. 1952, 20 July 1876, p. 230.
134. Pedder, 'Leading Points Regarding Revision Settlements'; NAI, RACD, LRSB, March 1875, 'A' proceedings 6–19, Keep-With Section.
135. DRCR, p. 105.
136. GOB to SOSI, No. 39, 17 December 1874, p. 1.
137. Lyon, too, argued that an increase that was fair on long-held land could not be reasonable on that taken up within the last 30 years, which was inevitably poorest and had had less time to benefit from low rates or increase in value. DRCR, Vol. II, p. 12.
138. Sumit Guha, *Agrarian Economy*, pp. 60–2.
139. This rule was re-affirmed in the Bombay Land Revenue Code Amendment Bill of 1879; See NAI, RAD, RB, December 1884, 'B' proceedings 84.
140. See Francis to Havelock, No. 1427, 7 October 1871; MSA, GOB, RD Vol. 57 of 1872, No. 134.
141. Havelock to GOB, No. 2831, 4 August 1874, pp. 105–6.
142. Francis to GOB, No. 1485, 19 September 1874; NAI, RACD, LRSB, March 1875, 'A' proceedings 6–19, p. 12.
143. Havelock to GOB, No. 1466, 20 April 1874, p. 7.
144. *Ibid.*, pp. 15–18.
145. *Ibid.*, p. 7.
146. Francis to GOB, No. 1485, 19 September 1874, p. 9.
147. *Ibid.*
148. *Ibid.*, p. 12.
149. GOB to SOSI, No. 39, 17 December 1874, p. 4.
150. Memorandum by Lord Northbrook, 10 November 1874; NAI, RACD, LRSB, March 1875, 'A' proceedings 6–19, Keep-With section.
151. GOI to SOSI, No. 30, 25 December 1874; NAI, RACD, LRSB, March 1875, 'A' proceedings 6–19, Keep-With Section.
152. GOI to GOB, No. 1094, 24 December 1874; *ibid.*
153. GOI to GOB, No. 539R, 15 May 1883; NAI, RAD, RB, May 1884, 'A' proceedings 37–8.
154. GOB to GOI, No. 6340, 27 August 1883.
155. GOB to SOSI, No. 39, 17 December 1874, p. 2.
156. Undated memorandum by B. H. Ellis, *c.*14 November 1874; NAI, RACD, LRSB, March 1875, 'A' proceedings 6–19, Keep-With section.
157. See NAI, RACD, LRSB, June 1876, 'B' proceedings 40–61.
158. SOSI to GOB, No. 34, 4 December 1879.
159. GOI to SOSI, No. 4, 17 March 1880; NAI HRAD, RB, March 1880, 'A' proceedings 82–3.
160. GOB to GOI, No. 634, 5 February 1880; *ibid.*
161. *Ibid.*
162. GOB to SOSI, No. 27, 24 December 1880; NAI, HRAD, RB, February 1881, 'A' proceedings 16–18.
163. SOSI to GOI, No. 73, 16 September 1880; NAI, HRAD, RB, November 1880, 'A' proceedings 32–3.
164. Memorandum by C. U. Aitchison, HRAD, 18 October 1880; *ibid.*
165. SOSI to GOB, No. 45, 11 November 1880; NAI, HRAD, RB, December 1880, 'A' proceedings 49. See also King to Robertson, No. 4161, 20 July 1880, p. 71.

166. King to Robertson, No. 4584, 22–5 July 1881, p. 142.
167. GOB to SOSI, No. 16, 24 July 1883; NAI, RAD, RB, August 1883, 'A' proceedings 33.
168. King to Robertson, No. 4584, 22–5 July 1881, pp. 139–41, plus reports and correspondence in NAI, RAD, RB, August 1881, 'B' proceedings 15 and September 1881, 'B' proceedings 19 and 20.
169. King to GOB, No. 772, 5 February 1881; NAI, RAD, RB, September 1881, 'B' proceedings 20, and undated memorandum by King; NAI, RAD, RB, August 1881, 'B' proceedings 15.
170. SOSI to GOB, No. 19, 19 October 1882, plus memoranda by Buck and Bayley; NAI, RAD, RB, July 1883, 'A' proceedings 47–9, Keep-With section.
171. GOB to SOSI, No. 16, 24 July 1883.
172. FCR (1880), Appendix I, Memorandum by Pedder, p. 157.
173. GOB Resolution, No. 6007, 21 October 1876; MSA, GOB, RD (Famine Branch) Vol. 58 of 1876–77, p. 3.
174. GOB Resolution, No. 6557, 15 November 1876; *ibid.*, p. 7.
175. Telegram from GOI to GOB, 16 January 1877; *ibid.*, p. 84.
176. Telegram from GOB to GOI, 3 February 1877; *ibid.*, p. 88.
177. GOB to GOI, No. 401, 22 January 1877; *ibid.*, p. 86.
178. GOI to GOB, No. 105, 29 January, 1877; *ibid.*, p. 87.
179. GOB to GOI, No. 774, 7 February 1877; *ibid.*, p. 69.
180. Telegram from GOI to GOB, 7 February 1877; *ibid.*, p. 73.
181. Temple to Salisbury, 17 November 1876; OIOC, Temple Collection.
182. John Strachey and Richard Strachey, *The Finances and Public Works of India from 1869 to 1881* (London, 1882), p. 27.
183. GOB to GOI, No. 606, 30 January 1877, Famine Vol. 58, p. 57.
184. See GOB resolutions and correspondence; *ibid.*, pp. 111–41.
185. GOB Resolution, No. 3846, 21 June 1877; *ibid.*, p. 151.
186. GOB to the Revenue Commissioners of all divisions, No. 5822, 25 September 1877; *ibid.*, p. 161.
187. Temple, 'The Famine', p. 115.
188. SOSI to GOB, No. 28, 30 October 1879; MSA, GOB, RD, Vol. 30 of 1880, pp. 4–5.
189. Jacomb to Robertson, No. A/4960, 19 July 1877, pp. 30–1.
190. Stewart to Robertson, No. 3195, 22–4 July 1878, p. 299. The unexplained difference was criticised in GOB Resolution No. 4280, 23 August 1878, AARA (1878), p. 577.
191. Stewart to Robertson, No. 3195, 22–4 July 1878, pp. 303–6.
192. King to Robertson, No. 3140, 19–23 July 1879, pp. 35–6. Once again, unexplained discrepancies were criticised in GOB Resolution No. 4672, 2 September 1879, AARA (1879), p. 284.
193. *Ibid.*
194. King to Robertson, No. 3140, 19–23 July 1879, p. 38.
195. King to Robertson, No. 4161, 20 July 1880, pp. 43–6.
196. GOB to SOSI, No. 17, 31 July 1880; NAI, RACD, RB, August 1880, 'B' proceedings 4.
197. King to Robertson, No. 4584, 22–5 July 1881, pp. 95–8.
198. Hamilton to King, No. 592, 9 July 1881, p. 304.
199. King to Robertson, No. 4584, 22–5 July 1881, p. 95.

200. Elphinston to Robertson, No. 5730, 20 July 1882, p. 138.
201. John Anding, Deputy District Collector and Magistrate, Ahmednagar, to Elphinston, No. 797, 10 July 1882, AARA (1882), p. 459. This includes the figures tabulated for Kopargaon, Rahuri and Newasa, and Rs 43,663 later sanctioned for Sangamner. See Elphinston to Robertson, No. 5730, 20 July 1882, p. 119.
202. GOB Resolution, No. 7272, 18 October 1882, p. 631.
203. Elphinston to Robertson, No. 5730, 20 July 1882, pp. 97–8.
204. GOB Resolution, No. 7272, 18 October 1882, p. 631.
205. Elphinston to Robertson, No. 5730, 20 July 1882, pp. 247–8.
206. *Ibid.*, pp. 30–7.
207. Anding to Elphinston, No. 797, 10 July 1882, p. 459.
208. *Ibid.*, p. 438.
209. *Ibid.*, p. 439.
210. Elphinston to Robertson, No. 5730, 20 July 1882, p. 133.
211. *Ibid.*, pp. 134–6, 139–40.
212. GOB Resolution, No. 7272, 18 October 1882, p. 631.
213. Deshpande to Elphinston, No. 599, 10 July 1882, p. 513.
214. GOB to GOI, No. 617, 4 February 1880; NAI, HRAD, RB, October 1880, 'A' proceedings 13–27.
215. GOB, RD Proceedings, 27 August 1883; MSA, GOB, RD (Famine Branch), Vol. 173, p. 155.
216. For example, GOB Resolution No. 3780, 26 July 1878, cited in King to Robertson, No. 3140, 19–23 July 1879, pp. 31–2.
217. SOSI to GOI, No. 73, 16 September 1880.
218. Memorandum by C. L. Tupper; NAI, HRAD, RB, November 1880, 'A' proceedings 32–3, Keep-With Section.
219. GOI Circular No. 56R, 12 October 1882, containing an extract of its proceedings; NAI, RAD, RB, October 1882, 'A' proceedings 53.
220. E. C. Ozanne, Secretary, Provincial Famine Code Committee, to GOB, No. 46, 22 September 1883; MSA, GOB, RD, Vol. 173, p. 277. Emphasis original.
221. GOB Resolution, No. 2619, 26 March 1884, para. 38.

5 Peasants and Relief Labour

1. Drèze, 'Famine Prevention', pp. 16–19; Rangasami, 'Failure', 1797–99.
2. See Ambirajan, *Classical Political Economy*, pp. 59–98; Hall-Matthews, 'Historical Roots', pp. 121–3.
3. See Ambirajan, 'Political Economy', 20–28; Hall-Matthews, 'Historical Roots', pp. 107–27.
4. See David Hall-Matthews, 'Perceptions of Famine: India in the 1870s', unpublished MA dissertation, School of Oriental and African Studies (London, 1994).
5. NNR, *Indu Prakash*, 9 April 1877.
6. NNR, *Dnyan Prakash*, 3 May 1877 and *Indu Prakash*, 7 May 1877.
7. Ground realities are intended to mean two, linked, things here: household socio-economic survival strategies and relief needs as identified by district officers. See Klein, 'When the Rains Failed', 185–214, in which he argues

that British famine policy tended generally to conflict with the former and fail to meet any of the latter but bare subsistence.
8. Digby argued that even this is not a sufficiently local level to see how the famine developed, pointing to the detailed evidence of the *Poona Sarvajanik Sabha* (PSS), whose agents claimed to 'live among and form part of the people overtaken by this calamity' whereas even local officers' views were 'naturally very deficient in the particular and intimate knowledge which ... is shut out to those whose range of vision necessarily embraces whole taluks or districts'. Digby, *The Famine*, Vol. I, p. 261. While there are indeed disadvantages in limiting the focus to British famine debates, it can be rejoindered that district officers in Ahmednagar, by running relief works, were to be found very nearly as close to the victims of famine as the elite urban Indians who made up the PSS. Moreover, while local bureaucrats could at least attempt to contest their orders from above, the PSS records had little influence on the colonial authorities (though they had an impact when published in Britain). See Hall-Matthews, 'Perceptions', pp. 19–21.
9. Havelock to GOB, No. 4811, 30 November 1870; MSA, GOB, RD, Vol. 23 of 1871, No. 70, citing D'Oyly. Emphasis added.
10. D'Oyly to Havelock, No. 1317, 21 October 1871; MSA, GOB, RD, Vol. 15 of 1871, No. 1225.
11. D'Oyly to Havelock, No. 2155, 11 November 1871; *ibid.*
12. GOI Resolution, January 1877, cited in Digby, *The Famine*, frontispiece.
13. GOB Resolution No. 208, 20 January 1871; MSA, GOB, RD, Vol. 23 of 1871, No. 70.
14. NNR, *Jame Jamsed*, 18 October 1876.
15. Arup Maharatna, *The Demography of Famines: An Indian Historical Perspective* (Delhi, 1996), p. 49.
16. See, for example, NNR, *Native Opinion*, 21 January 1877.
17. See Hall-Matthews, 'Historical Roots', pp. 107–13.
18. FCR (1880), Appendix I, pp. 109–20.
19. FCR (1867), part III, p. 174.
20. FCR (1880), Appendix I, p. 7.
21. *Ibid.*, p. 110.
22. *Ibid.*, p. 118.
23. FCR (1867), part III, p. 174.
24. John Strachey, 'Principles of Relief Measures by Government in Time of Famine in India', *The Friend of India*, April 1861. Brennan, 'Famine Code', p. 101, points out John Strachey's influence on his brother's appointment to the Famine Commission, suggesting that he would have been an ideal choice but for their relationship.
25. GOI Resolution No. 1967, 31 July 1876; MSA, GOB, RD, Vol. 126 of 1876, p. 8.
26. Quoted in GOB to GOI, No. 3362, 7 June 1876; MSA, GOB, RD, Famine Branch, Vol. 49 of 1876, p. 223.
27. Speech by Lord Lytton, *Famine Speeches*, p. 70.
28. RD to Temple, 16 January 1877, *Gazette of India*, Extraordinary issue.
29. Temple to Salisbury, 18 April 1877, Temple Collection. Emphasis original. The personalities and influence of the autocratic Stracheys and the dissenting but loyal Temple are discussed in detail in Hall-Matthews, 'Perceptions', pp. 6, 14–15, 29.

30. Major-General M. K. Kennedy, Secretary, GOB, 'Notes on the General Policy of the Government and on the Progress of the Measures Adopted to Relieve the Distress Caused by the Famine in the Deccan and Southern Maratha Country', 18 December 1876; MSA, GOB, General Department, Famine Volume, 1878; Digby, *The Famine*, Vol. I, pp. 296–314, who criticised this dispute as 'a game at cross purposes in high quarters'.
31. GRABP (1878), p. viii.
32. Robertson to GOB, No. 902, 22 February 1877; *ibid.*, p. 17.
33. Jacomb to Robertson, No. 879, 9/10 February 1877; Famine Vol. 24, p. 15.
34. Jacomb to Havelock, No. 3307, 17 November 1876, Famine Vol. 23, p. 109.
35. GOB Resolution No. 255 C.W.-975, 7 December 1876; *ibid.*, p. 130.
36. Jacomb to Robertson, No. 3713, 13/14 December 1876, Famine Vol. 23, p. 238.
37. Spry to Jacomb, No. 403, 30 December 1876, Famine Vol. 23, p. 401.
38. GOB Resolution No. 55 C.W.-132, 23 January 1877; *ibid.*, p. 405. Emphasis added.
39. Woodburn to Jacomb, No. 81, 5 February 1877, Famine Vol. 24, p. 451.
40. Report by Howard, Executive Engineer, Ahmednagar, cited in Superintending Engineer, CD to GOB, No. 1551, 1 May 1877; *ibid.*, p. 582.
41. Digby, *The Famine*, Vol. I, p. 384.
42. GRABP (1878), p. xiii.
43. FCR (1880), Appendix II, p. 50. The Bihar costs were calculated to include all works, land revenue remission and grain imports, without deductions for utility and specifically not for the value of leftover food stocks, for which Temple was especially pilloried.
44. Temple, 'The Famine', p. 111.
45. *Ibid.*, p. 117.
46. Speech by John Strachey, *Famine Speeches*, p. 8.
47. *Ibid.*, p. 4.
48. *Ibid.*, pp. 17, 18.
49. NNR, *Jame Jamsed*, 9 October 1877.
50. FCR (1880), Appendix I, p. 111.
51. Quoted in Havelock to GOB, No. 4811, 30 November 1870; MSA, GOB, GD, Vol. 23 of 1871, No. 70.
52. FCR (1880), 'Information and Evidence', chapter 1, question 26, answer by Robertson.
53. John Strachey, *Famine Speeches*, pp. 24–5.
54. GOB Resolution No. 1885, 12 April 1880, quoted in GOI to GOB, No. 467, 11 October 1880; NAI, GOI, HRAD, RB, 'A' proceedings 47.
55. Temple to John Strachey, 7 June 1878; OIOC, Strachey Collection.
56. Temple, 'Check and Supervision over Gratuitous Relief in Distressed Districts', *Minutes, Vol. I*, p. 50.
57. GOB Resolution No. 256 E-842, 16 April 1877 and Robertson to GOB, No. 2578, 30 May 1877, Famine Vol. 24, pp. 371, 647.
58. Temple, 'Check and Supervision' was written on 17 September 1877.
59. Jacomb to Robertson, No. 5563, 15 August 1877, Famine Vol. 25, p. 110.
60. Jacomb, Memorandum No. 2907, 23 October 1876, Famine Vol. 23, p. 35.
61. GOB Resolution No. 196 C.W.-849, 16 November 1876; *ibid.*, p. 41.
62. FCR (1880), part III, p. 176.
63. FCR (1880), part I, p. 24.

64. FCR (1880), part III, p. 176.
65. Ibid., p. 183.
66. GOB Resolution No. 315 C.W.-674, 27 March 1877, Famine Vol. 24, p. 243.
67. Ashburner to GOB, No. 1359, 10 April 1877; ibid., pp. 352–5.
68. Ibid., p. 354.
69. Sanjay Sharma, *Famine, Philanthropy*, pp. 166–8.
70. GOB Resolution No. 833, 11 November 1876; Famine Vol. 23, p. 25. See also Robertson to GOB, No. 688, 13 February 1877, ibid., p. 728.
71. GOB Resolution No. 833, 11 November 1876, p. 25.
72. *Poona Sarvajanik Sabha*, Famine narrative No. 4, cited in Digby, *The Famine*, Vol. I, p. 344.
73. Woodburn to Jacomb, No. 585, 23 August 1877, and Jacomb to Robertson, No. 5893, 28 August 1877, Famine Vol. 25, pp. 137–41.
74. Digby, *The Famine*, Vol. I, p. 268, citing correspondent Mr Curwen.
75. Robertson to GOB, No. 688, 13 February 1877, p. 730.
76. *Famine Relief Code – Bombay Presidency* (Poona, 1885), p. 8.
77. GOB Resolution No. 370 C.W.-826, 12 April 1877, Famine Vol. 24, p. 363. Emphasis added.
78. FCR (1880), Appendix I, p. 110.
79. T. G. Hewlett, Acting Sanitary Commissioner of Bombay, 'Rules for the information and guidance of those entrusted with the Sanitary arrangements in the Famine districts' and Hewlett to GOB No. 122/1176, 22 November 1876; MSA, GOB, RD (Famine Branch), Vol. 64 of 1876–77, p. 127.
80. Colonel C. J. Merriman, Chief Engineer for Irrigation, comments on Hewlett's 'Rules'; MSA, GOB, RD (Famine Branch), Vol. 64 of 1876–77, p. 126.
81. Hewlett to GOB, No. 122/1176, 22 November 1876, p. 128.
82. Hewlett, 'Rules', p. 122; Merriman, comments, p. 126.
83. GOB Resolution No. 3-F, 20 November 1876; ibid., p. 126. Emphasis original.
84. Hewlett, 'Rules', p. 125. Emphasis original.
85. Candy, for example, recorded cholera on the railway at Ghospuri in May 1877 and at a camp near Rahuri town in September. Quoted in Jacomb to Robertson, No. 3316, 16 May 1877, Famine Vol. 63, p. 544, and Candy to Stewart, No. 514, 11 July 1878, pp. 423–4.
86. Jacomb to Robertson, No. A/4960, 19 July 1877, p. 14.
87. Stewart to Robertson, No. 3195, 22–4 July 1878, p. 282.
88. Temple, 'The Famine', p. 107.
89. Klein, 'When the Rains Failed', 194.
90. David Arnold, 'Cholera Mortality in British India, 1817–1947', in Tim Dyson (ed.), *India's Historical Demography: Studies in Famine, Disease and Society* (London, 1989), pp. 261–84; Dyson, 'On the Demography', 14, 22; Lardinois, 'Famine, epidemics', 456–7; Whitcombe, 'Famine Mortality', 1172–5. All these authors refer to Madras Presidency, but relief conditions were likely to have been at least as insanitary in Bombay.
91. Spry to Jacomb, No. 349, 12 April 1877, Famine Vol. 24, pp. 467–8.
92. Temple to Salisbury, 22 February 1877. Temple complained somewhat implausibly that relief works in Madras contained many such people, blustering 'it is nothing short of an *abuse* that such men shd [sic] be in the receipt of state charity'. Emphasis original.

93. NNR, *Rast Goftar*, 28 January 1877.
94. Jacomb to Spry, No. 3796, 16/18 December 1876; *ibid.*, p. 656.
95. FCR (1880), 'Information and Evidence', chapter 3, question 4, reply by Hamilton.
96. FCR (1880), part III, p. 174.
97. Jacomb to Spry, No. 3796, 16/18 December 1876, p. 656.
98. Fforde to Jacomb, No. 1, 3 January 1877, Famine Vol. 23, p. 421.
99. Robertson to GOB, No. 192, 16 January 1877; *ibid.*, pp. 423–4.
100. GRABP (1878), p. ix.
101. *Nizam's Famine Report*, p. 67.
102. FCR (1880), 'Information and Evidence', chapter 3, question 36, reply by Wilkins.
103. *Ibid.*, chapter 3, question 4, reply by Hamilton.
104. FCR (1880), part III, p. 174.
105. Minute by Ashburner, unnumbered, 27 May 1881, Famine Vol. 173, p. 168.
106. See Himmelfarb, *The Idea*, pp. 79–83, 164.
107. Spry to Jacomb, No. 109, 7 February 1877, Famine Vol. 23, pp. 694–5.
108. GOB Resolution No. 33 E-103, 19 January 1877, Famine Vol. 62, pp. 351–2.
109. GOB Resolution No. 156 E-546, 10 March 1877, Famine Vol. 63, p. 69.
110. Howard to Bhasapa Erapa, Sub-overseer, No. 2003, 9 October 1876; *ibid.*, p. 59. Emphasis original.
111. Howard to all Subordinates in charge of works in the Ahmednagar Collectorate, No. 2231, 26 October 1876, Famine Vol. 23, p. 61. Emphasis original.
112. Jenkin Jones to GOB, No. 38D, 24 November 1876, Famine Vol. 23, pp. 149–50.
113. FCR (1880), 'Information and Evidence', chapter 3, question 10, reply by Hewlett.
114. Stephen Devereux, 'Transfers and Safety Nets', in Stephen Devereux and Simon Maxwell (eds), *Food Security in Sub-Saharan Africa* (London, 2001), p. 279. He distinguishes here between food-for-work and less 'stigmatising' cash-for-work, dominated by men in contemporary Malawi. However, low wages and the sliding scale make the distinction obsolete for Ahmednagar peasant households in 1877.
115. FCR (1880), cited in Jocelyn Kynch with Maureen Sibbons, 'Famine Relief, Piecework and Women Workers: Experiences in British India', in Helen O'Neill and John Toye (eds), *A World without Famine?*, p. 142.
116. Candy quoted in Robertson to GOB, No. 3729, 9/11 August 1877, Famine Vol. 63, p. 560.
117. Kynch, 'Famine relief', p. 143.
118. Candy quoted in Jacomb to Robertson, No. 3316, 16 May 1877, p. 543.
119. GOB to Robertson, No. 970F, 23 May 1877, Famine Vol. 63, p. 559.
120. Robertson to GOB, No. 3729, 9/11 August 1877, p. 561.
121. See G. B. Rodgers, 'Effects of Public Works on Rural Poverty: Some Case Studies from the Kesi Area of Bihar', *EPW*, VIII (4–6) (1973), 255–68.
122. Candy to Jacomb, unnumbered, 15 March 1877; *ibid.*, p. 230.
123. GOB Resolution No. 102 E-352, 14 February 1877, Famine Vol. 62, p. 427.
124. Lt. Knight, Sub-Engineer, to Howard, No. 38, 4 February 1877, Famine Vol. 25, p. 277.

125. Captain Alec Seton, Executive Engineer, to Superintending Engineer, CD, No. 283, 24 January 1877, Famine Vol. 63, pp. 31–2.
126. Minute by Sir Salar Jung, 20 Zikad 1296, *Nizam's* Famine Report, p. 3.
127. Hamilton to Jacomb, No. 311, 28 May 1877, Famine Vol. 24, p. 667.
128. Hamilton to Jacomb, No. 180, 16 March 1877, Famine Vol. 63, p. 228.
129. GOB Resolution No. 156, 10 March 1877, p. 69.
130. The debate on this subject in Madras, including Cornish's letters, is reprinted in FCR (1880), 'Compilation of Replies, Madras', Appendix B.
131. Temple to Salisbury, 10 March 1877.
132. Hewlett's timidity was criticised by the Poona newspaper *Dnyan Prakash*, but it recognised his difficulty after Cornish had been told 'to mind his own proper business'. NNR, 3 May 1877. When Hewlett sought to vindicate Temple by reporting that famine labourers were in 'remarkably good health', however, even the *Times of India* compared it to 'that kind of cheerful assertion ... which the compassionate physician gives on leaving the room of a dying patient'. Quoted in Digby, *The Famine*, Vol. II, p. 223.
133. See for example, Digby, *The Famine*, Vol. II, pp. 165–260; Bhatia, *Famines in India*, pp. 91–6; Hall-Matthews, 'Perceptions', pp. 7–9, 22–3.
134. GOB Resolution No. 166 C.W.-792, 4 November 1876, Famine Vol. 62, p. 3.
135. GOB Resolution No. 327 C.W.-1142, 29 December 1876; *ibid.*, p. 189.
136. GOB Resolution No. 268 C.W.-1038, 13 December 1876; *ibid.*, p. 121.
137. GRABP (1878), p. vii.
138. GOB Resolution, No. 33 E-103, 19 January 1877, p. 351.
139. *Ibid.*
140. GRABP (1877), p. 197.
141. Himmelfarb, *The Idea*, p. 181.
142. Jacomb to GOB, No. 5430, 9 August 1877, Famine Vol. 63, pp. 572–3.
143. Robertson to GOB, No. 1377, 22 March 1877, Famine Vol. 63, p. 196.
144. Jacomb to GOB, No. 5430, 9 August 1877, p. 574.
145. GOB Resolutions Nos. 89 E-1066, 19 December 1876 and 315 C.W.-1120, 27 December 1876, Famine Vol. 62, p. 157 and p. 167; Temple to GOB, 28 May 1877, Famine Vol. 63, p. 427.
146. FCR (1880), part III, p. 173.
147. GOB Resolution No. 317P, 10 August 1877, Famine Vol. 63, pp. 537–8.
148. Spry to Jacomb, No. 137, 16 February 1877; *ibid.*, p. 3 and GOB to Robertson, No. 427F, 1 March 1877, and Circular No. 521F, 15 March 1877; *ibid.*, pp. 6, 99.
149. Robertson to GOB, No. 1377, 22 March 1877, pp. 187–90.
150. Hamilton and Fforde jointly to Jacomb, No. 78, 30 January 1877; Famine Vol. 63, p. 737.
151. Jacomb to Robertson, No. 887, 9 February 1877; Famine Vol. 63, p. 718.
152. GOB Resolution No. 188 C.W.-391, 17 February 1877; Famine Vol. 63, p. 725.
153. Temple, 'The Famine', pp. 94–5.
154. *Bombay Gazette*, cited in Digby, *The Famine*, Vol. I, p. 370.
155. Cited in Digby, *The Famine*, Vol. I, p. 294.
156. NNR, *Dnyan Prakash*, 9 July 1877.
157. GRABP (1878), p. x.
158. FCR (1880), part III, p. 173.

159. Jacomb to Spry, Candy, Hamilton, Woodburn and Fforde, No. 706, 31 January 1877, Famine Vol. 63, p. 102.
160. Jacomb to Robertson, No. 1872, 19 March 1877; *ibid.*, p. 237.
161. Jacomb to Robertson, No. 1574, 7 March 1877; *ibid.*, pp. 104–5.
162. Hamilton to Jacomb, No. 131, 21 February 1877, Famine Vol. 24, p. 263.
163. *Ibid.*, p. 264.
164. Candy to Jacomb, unnumbered, 15 March 1877, p. 231.
165. Note by Wodehouse, 29 March 1877, Famine Vol. 63, pp. 205–8.
166. Minute by Ashburner, unnumbered, 27 May 1881, p. 167. Emphasis original.
167. FCR (1880), 'Information and Evidence, Bombay' and 'Compilation of Replies, Madras', chapter 3, question 6.
168. FCR (1880), 'Information and Evidence, Bombay', chapter 3, question 6, reply by Hamilton.
169. Dyson, 'On the Demography', pp. 22–5.
170. See Hall-Matthews, 'Historical Roots', pp. 113–17.
171. Cited in Famine Vol. 173, pp. 4–5.
172. Robertson, Collector of Dharwar, to Havelock, No. 3689, 18 November 1876, Famine Vol. 62, pp. 53–4.
173. Robertson to GOB, No. 1377, 22 March 1877, pp. 193–4.
174. *Ibid.*, pp. 195–6.
175. Robertson to GOB, No. 3729, 9/11 August 1877, p. 567.
176. See Hall-Matthews, 'Perceptions', pp. 14–15.
177. Sir Richard Temple, *India in 1880* (London, 1881), p. 340.
178. FCR (1880), Appendix II, p. 30.
179. Temple 'The Famine', p. 102.
180. Digby, *The Famine*, Vol. I, p. 345.
181. Jacomb to Robertson, No. 1872, 19 March 1877, p. 237.
182. Temple, 'The Famine', p. 102.
183. Digby, *The Famine*, Vol. I, pp. 340–4.
184. Spry to Jacomb, No. 116, 10 February 1877, Famine Vol. 24, p. 262.
185. Spry to Jacomb, No. 109, 7 February 1877, p. 694.
186. Candy to Jacomb, unnumbered, 15 March 1877, p. 231.
187. GOB Resolution No. 50 E-158, 15 January 1877, summarised in GOB to GOI, telegram, 12 February 1877, Famine Vol. 62, p. 437.
188. Fforde to Jacomb, No. 18, 2 February 1877, Famine Vol. 23, p. 733.
189. Candy to Jacomb, No. 53, 9 February 1877, Famine Vol. 62, p. 458.
190. GOI to GOB, telegram, 13 February 1877; *ibid.*, pp. 439–45.
191. Temple to Salisbury, 22 March 1877.
192. Temple, 'The Famine', p. 100.
193. *Ibid.*, p. 102.
194. FCR (1867), part II, p. 1.
195. FCR (1901), p. 2.
196. See J. B. Peile, 'Note on the Economic Condition of the Agricultural Population of India', FCR (1880), Appendix I, p. 162.
197. Minute by Ashburner, unnumbered, 27 May 1881, p. 168.
198. Raghunath Ramchandra to W. St. Clair, Executive Engineer, Ahmednagar, No. 59, 31 January 1877, Famine Vol. 62, p. 387.
199. Stanley Wolpert, *Tilak and Gokhale: Revolution and Reform in the Making of Modern India* (Delhi, 1989), p. 9.

200. FCR (1880), 'Information and Evidence', chapter 3, question 36, reply by Hewlett. The fear of famine migrants bringing disease to urban areas was very common. See David Arnold, *Famine: Social Crisis and Historical Change* (Oxford, 1988), p. 92 and Jim Masselos, 'Migration and Urban Identity: Bombay's Famine Refugees in the Nineteenth Century', in S. Patel and A. Thorner (eds), *Bombay: A Metaphor for Modern India* (Bombay, 1993), pp. 36–7.
201. Stewart to Robertson, No. 3196, 22–4 July 1878, p. 331.
202. See for example Woodburn to Jacomb, No. 481, 5 July 1877, pp. 176–8.
203. Arnold, 'Famine in Peasant Consciousness', pp. 87–93.
204. NNR, *Indu Prakash*, 16 October 1876.
205. GOB Resolution No. 6283, 12 October 1877, Famine Vol. 25, p. 247.
206. Jacomb to GOB, No. 3170, 9 September 1876, Famine Vol. 23, p. 45.
207. Jenkin Jones to GOB, No. 78D, 8 December 1876; *ibid.*, pp. 165–6.
208. *Nizam's* Famine Report, pp. 66–7.
209. *Ibid.*, p. 5.
210. *Ibid.*, p. 149.
211. *Ibid.*
212. Temple, Memorandum on the famine in the *Nizam's* Dominions, 12 January 1877; *ibid.*, p. 85.
213. Correspondence between Bennett Fitch, Woodburn, Jacomb, Robertson and GOB, 18 April–5 May 1877, Famine Vol. 24 , pp. 513–29.
214. Referred to in Jacomb to Robertson, No. 1153, 20/21 February 1877; *ibid.*, pp. 167–75.
215. Temple, 'The Famine', pp. 116–17; Lytton, *Famine Speeches*, pp. 73–4.
216. Spry to Jacomb, No. 54, 19 January 1877 and Hamilton to Jacomb, No. 113, 16 February 1877, Famine Vol. 24, pp. 169–73.
217. Woodburn to Jacomb, No. 35, 19 January 1877; *ibid.*, p. 174.
218. de Waal, *Famine That Kills*, p. 7.
219. Whitcombe, 'Famine Mortality', 1172.
220. Stewart to Robertson, No. 3195, 22–4 July 1878, p. 295.
221. Woodburn to Jacomb, No. 481, 5 July 1877, pp. 171–2.
222. For example, Anding to Elphinston, No. 797, 10 July 1882, p. 442.
223. King to himself, No. 164, 21 July 1879, p. 199.
224. Temple, 'The Famine', p. 95.
225. FCR (1880), part III, p. 168.
226. GOB Resolution No. 98P, 20 June 1877, Famine Vol. 64, p. 7.
227. FCR (1880), part III, p. 176.
228. Minute on the Famine by James Gibbs, 29 December 1877; MSA, GOB, General Department, Vol. 46 of 1878.
229. See King to himself, No. 164, 21 July 1879, p. 199.
230. Ashburner to GOB, No. 1359, 10 April 1877, p. 354.
231. Jacomb circular to Spry, Hamilton and Fforde, No. 3618, 7/8 December 1876, Famine Vol. 23, p. 654.
232. Hamilton to Jacomb, No. 180, 16 March 1877, p. 227.
233. Temple to Lytton, telegram, 7 May 1877, Famine Vol. 63, p. 399.
234. Woodburn to Jacomb, No. 645, 17 September 1877, Famine Vol. 25, p. 209.
235. Fforde to Jacomb, No. 113, 7 July 1877, Famine Vol. 64, p. 58.

236. *Ibid.*, pp. 58–9.
237. GOB Resolution No. 373P, 21 August 1877; *ibid.*, p. 61.
238. Hamilton to Jacomb, No. 300 of 1877, undated, Famine Vol. 24, pp. 643–4.
239. Jacomb to GOB, No. 5288, 2 August 1877, Famine Vol. 25, pp. 69–70.
240. Robertson to GOB, No. 1415, 27 March 1877, Famine Vol. 63, p. 240.
241. E. C. Ozanne, Secretary, Famine Code Committee, to GOB, No. 46, 22 September 1883, Famine Vol. 173, p. 277. Emphasis original.
242. Klein, 'When the Rains Failed', 204–7; K. Suresh Singh, 'The Famine Code: The context and Continuity', in Jean Floud and Amrita Rangasami (eds), *Famine and Society* (New Delhi, 1993), pp. 142–3.
243. Temple, 'The Famine', p. 107; 'Famine Mortality', *Minutes*, p. 91.
244. Sen, 'Famine Mortality', pp. 194–7.
245. Jacomb to GOB, No. 7701, 15 November 1877; MSA, GOB, RD (Famine Branch), Vol. 105 of 1876–77, p. 150.
246. Hewlett cited in Temple, 'Famine Mortality', pp. 90–1.
247. See Digby, *The Famine*, Vol. I, p. 377.
248. Temple, *India in 1880*, p. 336.
249. Temple, 'Relief inspection at/near Sholapur', 10 September 1877, *Minutes*, p. 47. Emphasis added.
250. FCR (1880), part I, p. 64.
251. *Ibid.*, p. 28.
252. FCR (1901), p. 2.
253. FCR (1880), 'Compilation of Replies, Madras', Vol. I, p. 23.
254. FCR (1880), part I, p. 28.
255. Temple, 'The Famine', p. 93.
256. Jacomb to GOB, No. 7901, 22 November 1877, Famine Vol. 25, p. 363.
257. Elphinston to Robertson, No. 5730, 20 July 1882, pp. 46–7.
258. Alexander de Waal and Rakiya Omaar, 'The Lessons of Famine', *Africa Report* (November 1992), 63.
259. Robertson to GOB, No. 3620, 3 August 1877, p. 264.
260. Temple, *India in 1880*, p. 340.
261. Temple, 'Relief Inspection', p. 50.
262. Candy to King, No. 362, 7–11 July 1879, p. 100.
263. For example, GOB Resolution, No. 33 E-103, 19 January 1877, p. 351.

Conclusion

1. For example, in Bhatia, *Famines in India*, pp. 14–21, 49–54, 131–3 and Hardiman, *Feeding the Baniya*, pp. 129–70.
2. Notably in McAlpin, *Subject to Famine*.
3. Strachey and Strachey, *Finances*, p. 159.
4. See Hall-Matthews, 'Historical roots', pp. 107–8, 117–19.
5. See Edkins, *Whose Hunger?*, p. xv.
6. Jodha, 'Famine', 1617–19.
7. McAlpin, *Subject to Famine*, 3–4, 191–4.
8. *Ibid.*, 185–90.

9. *Ibid.*, 198–202.
10. *Ibid.*, 199.
11. Boswell to Havelock, No. 2132, 20 July 1875, p. 470.
12. *Ibid.*, p. 471.
13. See James C. Scott, *Weapons of the Weak: Everyday Forms of Peasant Resistance* (New Haven, 1985).
14. GRABP (1876), p. xl.

Bibliography

Archive sources

Ahmednagar Collector's Office Record Room
Miscellaneous Correspondence, 1865, 1878, 1884.

Central Secretariat Library, New Delhi
Famine Speeches by Sir John Strachey, Sir A. Eden and Lord Lytton, 1878.

Maharashtra State Archives, Bombay
Government of Bombay, General Department Volumes, 1878.
Government of Bombay, Revenue Department, Famine Branch Volumes, 1876–77.
Government of Bombay, Revenue Department, Miscellaneous other Volumes, 1871–76 and 1880.
Government of Bombay, Revenue Department Publications, Annual *Jamabandi* Reports for Ahmednagar, 1876–84; Land and Sayer Revenue Reports for North and South Divisions, 1872–74, and for Central Division, 1880–81.
Government of Bombay, Revenue Department Volumes, Annual Administration Reports for Ahmednagar, 1874–82.

National Archives of India, New Delhi
Government of India, Revenue and Agriculture Department, General Branch Volumes, 1877.
Government of India, Revenue and Agriculture Department, Revenue Branch Volumes, 1873–84.
Government of India, Revenue and Agriculture Department, Tuccavee Branch Volumes, 1871–75.

Nehru Memorial Library, New Delhi
Poona Sarvajanik Sabha Quarterly Journal, 1880–82.

Oriental and India Office Collection, British Library, London
Compilations of Native Newspaper Reports, Bombay Presidency, 1876–78.
Correspondence of Sir Richard Temple with Lord Salisbury, 1874–77, Temple Collection.
Correspondence of Sir Richard Temple with Sir John Strachey, 1878, Strachey Collection.

Government publications

Ahmadnagar Gazetteer, Revised edition (Government Central Press, Poona, 1976).
Bombay Famine Relief Code (Government Central Press, Poona, 1885).
Bombay City Gazetteer, Vol. III (Government Central Press, Bombay, 1910).
Bombay Gazetteer, Vol. XVII: Ahmadnagar, edited by James Campbell (Government Central Press, Poona, 1884).
Deccan Riots Commission Report, 2 Volumes and 2 Appendices (Government Central Press, Poona, 1875).
Famine Commission Report (Eyre & Spottiswoode, London, 1867).
Famine Commission Report, 3 Volumes, 2 Appendices and Collected Evidence from Bombay and Madras Presidencies (Eyre & Spottiswoode, London, 1880–85).
Famine Commission Report (Government Central Press, Calcutta, 1901).
General Reports on the Administration of Bombay Presidency (Government Central Press, Poona, 1870–78).
Minutes by His Excellency Sir Richard Temple, Governor of Bombay, 2 Volumes (Government Central Press, Poona, 1877–80).
Report on the History of the Famine in His Highness the Nizam's Dominions (Bombay, 1879).

Secondary sources

Agarwal, Bina, 'Social Security and the Family: Coping with Seasonality and Calamity in Rural India', in Ahmad, E., Dreze, J., Hills, J. and Sen, A. (eds), *Social Security in Developing Countries* (Clarendon, Oxford, 1991), pp. 171–244.
Alamgir, Mohiuddin, *Famine in South Asia* (Velgeschlager, Gunn & Hain, Cambridge, MA, 1980).
Alavi, Hamza, 'Peasants and Revolution', in Gough, Kathleen and Sharma, H. P. (eds), *Imperialism and Revolution in South Asia* (Monthly Review Press, New York, 1973), pp. 291–337.
Ambirajan, Srinivasa, 'Political Economy and Indian Famines', *South Asia*, 1 (2) (1971), 19–28.
——, 'Malthusian Population Theory and Indian Famine Policy in the Nineteenth Century', *Population Studies*, 30 (1) (1976), 5–14.
——, *Classical Political Economy and British Policy in India* (Cambridge University Press, Cambridge, 1978).
Anderson, F. G. H., *Facts and Fallacies about the Bombay Land Revenue System* (Self-published, Poona, 1929).
Arnold, David, 'Looting, Grain Riots and Government Policy in South India 1918', *Past and Present*, 84 (1979), 111–45.
——, 'Famine in Peasant Consciousness and Peasant Action: Madras 1876-8', in Guha, Ranajit (ed.), *Subaltern Studies III* (Oxford University Press, Delhi, 1984), pp. 62–115.
——, *Famine: Social Crisis and Historical Change* (Basil Blackwell, Oxford, 1988).
——, 'Cholera Mortality in British India, 1817–1947', in Dyson, Tim (ed.), *India's Historical Demography: Studies in Famine, Disease and Society* (Curzon Press, London, 1989), pp. 261–84.

Baden-Powell, B. H., *The Land Systems of British India, Volume 1* (Oxford University Press, Oxford, 1892).
Ballhatchet, Kenneth, *Social Policy and Social Change in Western India, 1817–1830* (Oxford University Press, London, 1957).
Banaji, Jairus, 'Capitalist Domination and the Small Peasantry: Deccan Districts in the Late Nineteenth Century', *Economic and Political Weekly*, XII (33 and 34) (1977), 1375–1404.
Bayly, C. A., *Indian Society and the Making of the British Empire* (Cambridge University Press, Cambridge, 1988).
Bhatia, B. M., *Famines in India: A Study in Some Aspects of the Economic History of India with Special Reference to Food Problem, 1860–1990* (3rd edn, Konark, Delhi, 1991).
Bhattacharya, Neeladri, 'Agricultural Labour and Production: Central and South-East Punjab', in K. N. Raj, Bhattacharya, Neeladri, Guha, Sumit and Padhi, Sakti (eds.), *Essays on the Commercialisation of Indian Agriculture* (Oxford University Press, Delhi, 1985), pp. 105–62.
Boserup, Ester, *The Conditions of Agricultural Growth* (George Allen and Unwin, London, 1965).
Brennan, Lance, 'The Development of the Indian Famine Code', in Currey, Bruce and Hugo, Graeme (eds), *Famine as a Geographical Phenomenon* (Reidel, Dordrecht, 1984), pp. 91–111.
——, 'Government Famine Relief in Bengal, 1943', *Journal of Asian Studies*, 47 (1988), 542–67.
Buchanan-Smith, Margaret and Davies, Susanna, *Famine Early Warning and Response: The Missing Link* (Intermediate Technology, London, 1995).
Byres, Terry, 'Land Reform, Industrialization and the Marketed Surplus in India: An Essay on the Power of Rural Bias', in Lehmann, David (ed.), *Agrarian Reform and Agrarian Reformism: Studies of Peru, Chile, China and India* (Faber, London, 1974), pp. 221–61.
Catanach, Ian, *Rural Credit in Western India, 1875–1930: Rural Credit and the Co-operative Movement in the Bombay Presidency* (University of California Press, Berkeley, CA, 1970).
——, 'Agrarian Disturbances in Nineteenth Century India', in Hardiman, David (ed.), *Peasant Resistance in India, 1858–1914* (Oxford University Press, Delhi, 1992), pp. 184–203.
Charlesworth, Neil, 'Rich Peasants and Poor Peasants in Late Nineteenth-Century Maharashtra', in Dewey, Clive and Hopkins, A. G. (eds), *The Imperial Impact: Studies in the Economic History of Africa and India* (Athlone Press, London, 1978), pp. 97–113.
——, *Peasants and Imperial Rule: Agriculture and Agrarian Society in the Bombay Presidency, 1850–1935* (Cambridge University Press, Cambridge, 1985).
——, 'The Myth of the Deccan Riots', in Hardiman, David (ed.), *Peasant Resistance in India, 1858–1914* (Oxford University Press, Delhi, 1992), pp. 204–26.
Clay, Edward and Schaffer, Bernard, *Room for Manoeuvre: An Exploration of Public Policy in Agriculture and Rural Development* (Heinemann, London, 1984).
Currey, Bruce, 'Coping with Complexity in Food Crisis Management', in Currey, Bruce and Hugo, Graeme (eds), *Famine as a Geographical Phenomenon* (Reidel, Dordrecht, 1984), pp. 183–202.
Davis, Mike, *Late Victorian Holocausts: El Nino Famines and the Making of the Third World* (Verso, London, 2001).

Bibliography

Devereux, Stephen, Howe, Paul and Biong Deng, Luka, 'Introduction', *IDS Bulletin*, 33 (4) (2002).

Devereux, Stephen, *Theories of Famine* (Harvester Wheatsheaf, Hemel Hempstead, 1993).

——, 'Transfers and Safety Nets', in Devereux, Stephen and Maxwell, Simon, *Food Security in Sub-Saharan Africa* (ITDG Publishing, London, 2001), pp. 267–93.

de Waal, Alexander, *Famine that Kills: Darfur, Sudan, 1984–85* (Clarendon, Oxford, 1989).

——, *Famine Crimes: Politics and the Disaster Relief Industry in Africa* (James Currey, Oxford, 1997).

de Waal, Alexander and Omaar, Rakiya, 'The Lessons of Famine', *Africa Report* (November 1992).

Dewey, Clive, '*Patwari* and *Chaukidar*: Subordinate Officials and the Reliability of India's Agricultural Statistics', in Dewey, Clive and Hopkins, A. G. (eds), *The Imperial Impact: Studies in the Economic History of Africa and India* (Athlone Press, London, 1978), pp. 280–314.

——, *Anglo-Indian Attitudes: The Mind of the Indian Civil Service* (Hambledon, London, 1993).

Digby, William, *The Famine Campaign in South India*, 2 Volumes (Longman, Green, London, 1878).

Drèze, Jean, 'Famine Prevention in India', in Drèze, Jean and Sen, Amartya (eds), *The Political Economy of Hunger, Vol. II, Famine Prevention* (Clarendon, Oxford, 1990), pp. 13–122.

Drèze, Jean and Sen, Amartya, *Hunger and Public Action* (Clarendon, Oxford, 1989).

Dutt, Romesh, *Open Letters to Lord Curzon on Famines and Land Assessments in India* (Kegan Paul, Trench, Trubner, London, 1900).

——, 'The Indian Land Question', in Romesh Dutt (ed.), *Land Problems in India* (G. A. Natesan, Madras, 1903).

——, *The Economic History of India, Vol. I* (Kegan Paul, Trench, Trubner, London, 1902).

Dyson, Tim (ed.), *India's Historical Demography: Studies in Famine, Disease and Society* (Curzon Press, London, 1989).

——, 'On the Demography of South Asian Famines, Part I', *Population Studies*, 45 (1991), 5–25.

Edkins, Jenny, *Whose Hunger? Concepts of Famine, Practices of Aid* (University of Minnesota Press, Minneapolis, MN, 2000).

Floud, Jean and Rangasami, Amrita (eds), *Famine and Society* (Indian Law Institute, New Delhi, 1993).

Frykenberg, R. E., *Guntur District, 1788–1848: A History of Local Influence and Central Authority in South India* (Clarendon, Oxford, 1965).

Greenough, Paul, *Prosperity and Misery in Modern Bengal: The Famine of 1943–44* (Oxford University Press, Oxford, 1982).

Guha, Ranajit, 'On Some Aspects of the Historiography of Colonial India', in Ranajit, Guha (ed.), *Subaltern Studies I* (Oxford University Press, Delhi, 1982), pp. 1–9.

Guha, Sumit, *The Agrarian Economy of the Bombay Deccan, 1818–1941* (Oxford University Press, Oxford, 1985).

———, 'The Land Market in Upland Maharashtra', *Indian Economic and Social History Review*, 24 (2) (1987), 117–44.
Hacking, Ian, *The Taming of Chance* (Cambridge University Press, Cambridge, 1990).
Hall-Matthews, David, 'Perceptions of Famine: India in the 1870s', unpublished MA dissertation (School of Oriental and African Studies, London, 1994).
———, 'The Historical Roots of Famine Relief Paradigms', in O'Neill, Helen and Toye, John (eds), *A World without Famine? New Approaches to Aid and Development* (Macmillan, Basingstoke, 1998), pp. 107–27.
Hardiman, David (ed.), *Peasant Resistance in India, 1858–1914* (Oxford University Press, Delhi, 1992).
———, *Feeding the Baniya: Peasants and Usurers in Western India* (Oxford University Press, Delhi, 1996).
Harnetty, Peter, 'Cotton Exports and Indian Agriculture, 1861–1870', *Economic History Review*, 24 (3) (1971), 414–29.
Haynes, Douglas, 'Market Formation in Khandesh, c.1820–1920', *Indian Economic and Social History Review*, 36 (3) (1999), 275–302.
Himmelfarb, Gertrude, *The Idea of Poverty: England in the Early Industrial Age* (Faber, London, 1984).
Hurd, John, 'Railways', in Kumar, Dharma (ed.), *The Cambridge Economic History of India, Vol. 2* (Cambridge University Press, Cambridge, 1983), pp. 737–61.
Jodha, N. S., 'Famine and Famine Policies: Some Empirical Evidence', *Economic and Political Weekly*, 10 (41) (1975), 1609–23.
Keen, David, *The Benefits of Famine: A Political Economy of Famine and Relief in Southwestern Sudan, 1983–1989* (Princeton University Press, Princeton, NJ, 1994).
Kinealy, Christine, *A Death-Dealing Famine: The Great Hunger in Ireland* (Pluto, London, 1997).
Klein, Ira, 'When the Rains Failed: Famine, Relief and Mortality in British India', *Indian Economic and Social History Review*, 21 (2) (1984), 185–214.
Kumar, Dharma, *Land and Caste in South India: Agricultural Labour in the Madras Presidency in the Nineteenth Century* (Cambridge University Press, Cambridge, 1965).
Kumar, Ravinder, *Western India in the Nineteenth Century: A Study in the Social History of Maharashtra* (Routledge and Kegan Paul, London, 1968).
———, 'The Deccan Riots of 1875', in Hardiman, David (ed.), *Peasant Resistance in India, 1858–1914* (Oxford University Press, Delhi, 1992), pp. 153–83.
Kynch, Jocelyn and Sibbons, Maureen, 'Famine Relief, Piecework and Women Workers: Experiences in British India', in O'Neill, Helen and Toye, John (eds), *A World without Famine? New Approaches to Aid and Development* (Macmillan, Basingstoke, 1998), pp. 128–57.
Lardinois, Roland, 'Famine, Epidemics and Mortality in South India: A Reappraisal of the Demographic Crisis of 1876–1878', *Economic and Political Weekly*, XX (11) (1985), 454–65.
Maharatna, Arup, *The Demography of Famines: An Indian Historical Perspective* (Oxford University Press, Delhi, 1996).
Mann, Harold, *Rainfall and Famine: A Study of Rainfall in the Bombay Deccan, 1865–1938* (Indian Society of Agricultural Economics, Bombay, 1955).

Masselos, Jim, 'Migration and Urban Identity: Bombay's Famine Refugees in the Nineteenth Century', in Patel, S. and Thorner, A. (eds), *Bombay: A Metaphor for Modern India* (Oxford University Press, Bombay, 1993), pp. 25–58.

Marx, Karl, *Capital, Vol. III*, trans. David Fernbach (Penguin, Harmondsworth, 1981).

McAlpin, Michelle, 'Railroads, Cultivation Patterns and Foodgrains Availability: India 1860–1900', *Indian Economic and Social History Review*, 12 (1) (1975), 43–60.

——, *Subject to Famine: Food Crises and Economic Change in Western India, 1860–1920* (Princeton University Press, Princeton, NJ, 1983).

Metcalf, Tom, *The Aftermath of Revolt: India, 1857–1870* (Princeton University Press, Princeton, NJ, 1964).

Morris, Morris D., 'Economic Change and Agriculture in Nineteenth Century India', *Indian Economic and Social History Review*, 3 (2) (1966), 185–209.

Nand, Brahma, 'The Deccan Peasant Uprising, 1875', unpublished M.Phil thesis (Jawaharlal Nehru University, New Delhi, 1980).

O'Hanlon, Rosalind, *Caste, Conflict and Ideology: Mahatma Jotirao Phule and Low Caste Protest in Nineteenth-Century Western India* (Cambridge University Press, Cambridge, 1985).

Perlin, Frank, 'Of White Whale and Countrymen in the Eighteenth Century Maratha Deccan: Extended Class Relations, Rights and the Problem of Rural Autonomy under the Old Regime', *Journal of Peasant Studies*, 5 (2) (1978), 172–237.

Rabitoy, Neil, 'System v Expediency: The Reality of Land Revenue Administration in the Bombay Presidency, 1812–20', *Modern Asian Studies*, 9 (4) (1975), 529–46.

Raj, K. N., Bhattacharya, Neeladri, Guha, Sumit and Padhi, Sakti (eds), *Essays on the Commercialisation of Indian Agriculture* (Oxford University Press, Delhi, 1985).

Rangarajan, Mahesh, *Fencing the Forest: Conservation and Ecological Change in India's Central Provinces, 1860–1914* (Oxford University Press, Delhi, 1996).

Rangasami, Amrita, ' "Failure of Exchange Entitlements" Theory of Famine: A Response', *Economic and Political Weekly*, XX (41 and 42) (1985), 1747–52, 1797–1801.

Ravallion, Martin, *Markets and Famines* (Clarendon, Oxford, 1987).

Robb, Peter, 'Introduction. Land and Society: The British "Transformation" in India', in Robb, Peter (ed.), *Rural India: Land, Power and Society under British Rule* (Oxford University Press, Delhi, 1983), pp. 1–23.

Rodgers, G. B., 'Effects of Public Works on Rural Poverty: Some Case Studies from the Kosi Area of Bihar', *Economic and Political Weekly*, VIII (4–6) (1973), 255–68.

Scott, James C., *The Moral Economy of the Peasant: Rebellion and Subsistence in Southeast Asia* (Yale University Press, New Haven, CT, 1976).

——, *Weapons of the Weak: Everyday Forms of Peasant Resistance* (Yale University Press, New Haven, CT, 1985).

——, *Seeing Like a State: How Certain Schemes to Improve the Human Condition Have Failed* (Yale University Press, New Haven, CT, 1998).

Sen, Amartya K., 'Famine Mortality: A study of the Bengal Famine of 1943', in Hobsbawm, Eric, Kula, Witold, Mitra, Ashok, Raj, K. N. and Sachs, Ignacy (eds), *Peasants in History: Essays in Honour of Daniel Thorner* (Oxford University Press, Calcutta, 1980), pp. 194–220.

——, *Poverty and Famines: An Essay on Entitlement and Deprivation* (Clarendon, Oxford, 1981).
Sharma, Sanjay, *Famine, Philanthropy and the Colonial State: North India in the Early Nineteenth Century* (Oxford University Press, Delhi, 2001).
Singh, K. Suresh, 'The Famine Code: The Context and Continuity', in Floud, Jean and Rangasami, Amrita (eds), *Famine and Society* (Indian Law Institute, New Delhi, 1993), pp. 139–61.
Sivaswamy, K. G., *Legislative Protection and Relief of Agriculturist Debtors in India* (Gokhale Institute of Politics and Economics, Poona, 1939).
Spangenberg, Bradford, *British Bureaucracy in India: Status, Policy and the ICS in the Late Nineteenth Century* (Oxford University Press, Delhi, 1975).
Srivastava, H. S., *The History of Indian Famines and Development of Famine Policy* (Sri Ram Mehra, Agra, 1968).
Stokes, Eric, *The English Utilitarians and India* (Clarendon, Oxford, 1959).
——, *The Peasant and the Raj: Studies in Agrarian Society and Peasant Rebellion in Colonial India* (Cambridge University Press, Cambridge, 1978).
Strachey, John, 'Principles of Relief Measures by Government in Time of Famine in India', *The Friend of India*, April 1861.
Strachey, John and Strachey, Richard, *The Finances and Public Works of India from 1869 to 1881* (Kegan & Paul, London, 1882).
Temple, Richard, *India in 1880* (John Murray, London, 1881).
Thompson, E. P., 'The Moral Economy of the English Crowd in the Eighteenth Century', *Past and Present*, 50 (1971), 76–136.
Washbrook, David, 'Law, State and Agrarian Society in Colonial India', *Modern Asian Studies*, 15 (1) (1981), 649–721.
——, 'Progress and Problems: South Asian Economic and Social History c.1720–1860', *Modern Asian Studies*, 22 (1) (1988), 57–96.
——, 'The Commercialization of Agriculture in Colonial India: Production, Subsistence and Reproduction in the "Dry South", c.1870–1930', *Modern Asian Studies*, 28 (1) (1994), 129–64.
Wedderburn, William, *A Permanent Settlement for the Deccan* (Self-published, Bombay, 1880).
Whitcombe, Elizabeth, 'Famine Mortality', *Economic and Political Weekly*, XVIII (23) (5 June 1993), 1169–79.
Wink, Andre, *Land and Sovereignty in India: Agrarian Society and Politics under the Eighteenth-Century Maratha Svarajya* (Cambridge University Press, Cambridge, 1986).
Wolpert, Stanley, *Tilak and Gokhale: Revolution and Reform in the Making of Modern India* (Oxford University Press, Delhi, 1989).

Index

Page numbers in bold are for tables; those followed by an n are for endnotes.

agricultural bank, proposal for, 93, 123–6
Ahmednagar district
  colonial perceptions of, 19–22, 39–52, 58, 63, 81–2, 92–6, 123–4, 139, 147, 154, 200
  soil, low quality of, 19, 25, 27, 39, 48, 50–1, 56, 60, 62, 89, 113, 115, 147, 149, 168, 212

Bhatia, B. M., 85, 89, 120, 130, 136, 222 (n8), 223 (n16), 252 (n133), 255 (n1)
Bhils, 74–5, 106, 202, 209, 238 (n127)
Boswell, Henry, Collector of Ahmednagar, 13, 15, 28, 31–2, 35, 47, 49–52, 61–2, 65, 71, 77, 94, 101, 103–5, 107–8, 112, 118, 150–1, 157, 193, 218–9, 221

cattle, *see* livestock
Charlesworth, Neil, 4, 17, 32, 74, 78, 93, 96, 100–2, 104, 110, 113, 136, 237 (n77)
colonial administration, *see* state
coping strategies, 3, 220, 247–8 (n7)
  crime, 4, 198, **199**
  migration, 8, 42, 74, 78, 141, 170, 197–204 (**201**), 206–11
  sale of productive assets and jewellery, 78, 81, 173, 184, 203
cotton
  markets in, 58–63, 67–8, 72, 76, 86, 89–90
  production of, 4, 36–9 (**37**, **38**), 50, 57, 59–63 (**60**), 67–8, 213–14
Cranbrook, Lord, Secretary of State for India, 46, 85, 107, 134, 138, 154–5, 161–3, 165–6, 230 (n4)
credit, 92–127
  decline in availability of, 93, 95, 100–2, 104–5, 108–14, 127, 139, 148, 161, 167, 173, 190, 212, 214–15, 217
  legislation to reduce, 79, 91, 92–3, 97–104, 107–11, 212, 214–15, 219
  markets in, 3, 9, 59, 69–70, 79–80, 89, 95–8, 102, 115, 214, 218
  role of civil courts in, 23, 69–70, 93, 95, 97–103, 108–12, 124, 126, 220
  use of, as insurance against risk, 4–5, 17, 81, 93, 96, 102–4, 109–10, 113, 203
  see also takavi loans
crime, *see* coping strategies
cultivation, levels of, 27–32 (**31**), 37–9
  decline in, 8, 17, 30–3, 44–8, 56, 67, 80, 152, 158, 162, 214, 220
  overexpansion of, 3, 27, 32–3, 47, 66, 95, 130, 135, 152, 216
  see also peasants – cultivation strategies; landholdings

Davis, Mike, 7–8
Deccan Agriculturists' Relief Act (1879), 16–17, 23, 39, 48, 70, 79, 92, 96, 100–4, 107–12, 117, 121, 126, 138, 148, 155, 212, 215
Deccan Riots, 5, 14, 16–17, 70, 92, 102, 104–7, 114, 127, 137, 141, 151, 199, 212, 220–1
Deccan Riots Commission, 16, 19, 22, 25, 36, 81, 93–111, 113, 126–7, 138, 151, 214
development agenda, *see* Government of Bombay – agrarian policy
Devereux, Stephen, 3–4, 7, 186, 224 (n24)
de Waal, Alex, 1, 106, 201, 232 (n80), 255 (n258)

Index 265

Digby, Sir William, 14, 196, 248 (n8), 249 (n30), 252 (n133)
disease, 4, 107, 182–3, 185, 194, 198, 204–6, 209
drought, *see* rainfall, unreliability of
Dutt, Romesh, 136–7, 139, 147–8, 218

economic conditions
  effects of American civil war, 42, 57, 59–63, 67–9, 89, 95
  global recession (1873–79), 3–4, 57–8, 67–9, 80, 89, 100, 111, 116, 130, 143, 148, 173, 190, 213, 215
Elphinston, John, Collector of Ahmednagar, 12–13, 42, 48–9, 75, 80, 110, 115, 121, 140, 164, 208, 221
employment opportunities, *see* labour markets
entitlements failure, 1, 4, 57, 66, 68, 83, 168, 213, 216

famine, colonial conceptions of, 2, 41–2, 45, 48, 157, 161, 168–71, 216, 233 (n100)
famine codes, 2, 12, 82, 88, 114, 157, 170, 172–3, 182, 184, 193–4, 204, 210, 215–16
Famine Commission (1880), 2, 9, 14, 16, 19–20, 30, 39, 41, 49, 51, 76, 85, 92, 94, 102, 105, 112, 115, 118–20, 122–4, 130, 135, 140, 172–4, 178, 180, 184, 186, 192–4, 202–3, 205, 210, 214–16
famine crisis (1876–78), 7–8, 16, 33, 35, 39–45, 52, 56, 57, 65–6, 78, 80–6, 89–92, 95, 100–1, 104–5, 108, 111–12, 118–24, 127–30, 132–4, 136, 143, 154–5, 157–220 (**171**)
famine mortality, 40, 57, 90–1, 168, 170–1, 178, 182, 192, 194, 196–8, 201–2, 204–10 (**206–8**), 219
famine as long-term process, 1–10, 12–16, 34, 45–8, 56, 89–91, 95, 102, 104–7, 111, 128–9, 141, 143, 150–1, 155–7, 161, 165–7, 185, 192, 202, 207–21
famine relief, *see* relief works; gratuitous relief
foodgrain markets, 3–4, 59, 79
  extent of commercialisation of, 4, 6–9, 17, 57–8, 61–82, 84, 89–91, 135, 213, 218
  falling and fluctuating prices in, 8, 28, 32, 56–7, 62, 64–71 (**64, 65**), 80–2, **83**– 6, 89–90, 104, 116, 129–30, 136, 145–6, 149, 167, 188–90, 212–3, 215, 217
  lack of integration of, 4, 17, 59–70 (**69, 70**), 75–8, 89, 212–4
  over-reliance on, during famine, 17, 72, 75, 82–8, 176, 209, 216
foodgrain production, 4, 33, 36–51 (**37, 38**), 56, 58, 61, 68, 79–81, 94–5, 138, 152, 164–5, 185, 217
forest reserves, 28, 51–3

Government of Bombay
  agrarian policy, 2–3, 5–9, 14–17, 19–21, 28–9, 32–3, 48–56, 57–68, 71, 81–2, 88–93, 100, 108, 112–28, 133–6, 141, 143, 152–8, 169–70, 178, 213–19
  famine policy, 1–4, 9–10, 12–18, 82–8, 116, 118–20, 128, 159–61, 168–211, 215–20
  inter-departmental rivalry, 11, 51–3, 129, 145, 147, 149–53, 156, 182–3, 186–7, 210–11
  internal hierarchy, disputes within, 10–14, 45–51, 76–7, 81, 87–8, 93, 121, 127, 129, 147–9, 150–3, 164, 167, 169–70, 175–6, 182, 185–6, 192–7, 202–4, 210, 216, 219
  relationship with Government of India, 11, 16, 55, 67, 115, 118–19, 122, 125, 129, 132–3, 146, 149–50, 153–7, 159–60, 166–7, 173–5, 178–9, 197, 219
Government of India, 10–11, 46, 54–5, 72, 94, 103, 106–7, 115–19, 122–3, 125–6, 130–1, 134, 137, 139–40, 143, 145–8,

266 Index

Government of India – *continued*
  154–7, 159–60, 166–7, 170–4,
  215, 217
Government of Madras, 82, 88,
  117–18, 133, 160, 174, 188,
  191–3, 200
gratuitous relief, 179–80, 185, 196,
  204–5
Guha, Sumit, 3–4, 29, 33, 95–7, 135,
  152, 216, 230 (n4)

Hartington, Lord, Secretary of State
  for India, 145–6, 149–50, 155–7
Havelock, William, Revenue
  Commissioner, 12–13, 33, 40,
  51, 80, 93–4, 102, 141, 147, 149,
  151–3, 156
Hewlett, T. G., Sanitary
  Commissioner, Government of
  Bombay, 182, 186, 188, 198, 205
Hume, A. O., 50, 55

indebtedness, 4, 17, 23, 56, 66, 76,
  79–80, 92–127, 159–60, 164–5,
  214, 220
  see also credit
irrigation, 27, 53–6, 61, 79, 114–18,
  122, 145, 152, 176, 179, 181

Khandesh district, 49–50, 54, 61,
  109, 119
Kimberley, Lord, Secretary of State
  for India, 122, 125–6, 143

labour markets, 6, 8, 17–18,
  22–3, 75, 81, 116, 168–9,
  175, 188–90 (**189**), 210–2,
  214, 217
laissez-faire, 6–7, 20, 57–9, 72, 75–6,
  81–91, 106, 108, 112–13, 127,
  132, 168–9, 171–2, 198, 200,
  209, 211, 213, 216, 219
  see also foodgrain markets –
  commercialisation of
landholdings
  abandonment of, 3, 17, 28–30
    (**29**), 44, 52, 104, 130, 141–4,
    158, 161–3, 165, 199–200, 210,
    217–18, 220

insecurity of tenure on, 8–9, 21–5,
  28–30, 100–5, 107–9, 113, 117,
  141, 214
sizes of, 25–8 (**26, 27**)
transfer to moneylenders, 23–5, 28,
  30, 78, 100–4, 107–8, 111,
  144, 212, 217
  see also cultivation, levels of
landless labourers, 6, 8, 22–3, 90,
  152, 168, 179, 199, 201–3, 210,
  212, 216
land revenue
  30-year revision of, 5, 11, 16–17,
    45, 56, 62, 68, 80–1, 91,
    104–5, 110, 115, 121, 125,
    129–31, 133–7, 139, 141,
    143–57, 161, 164, 167, 212,
    214–15, 217–18, 226 (n54)
  calls for lighter system of, 46, 68,
    81, 105–6, 129–31, 134,
    146–51, 155–6, 166–7, 212
  collection of, 3, 5, 14, 66–7, 110,
    113–14, 125, 129–30,
    139–40, 143, 157–67, 173,
    214, 217
  incidence of, 7, 33, 53, 55, 79,
    81–2, 128–31 (**131**), 133–8,
    147, 179, 192, 214, 217–18
  inflexibility of, 17, 28–30, 94,
    111, 128–9, 138–43, 148,
    166, 212–5, 217,
    219–20
  links to credit markets, 93–5, 98,
    104–7, 110–14, 128–9, 136,
    138, 151, 159–60, 167, 215
  penalties for non-payment of,
    29–30, 104, 107, 113, 141–144
    (**142**), 159–60, 163, 165,
    199–200
  ryotwari system of, 5, 17, 29,
    128–9, 134–9, 143–8, 166–7,
    196, 198, 213–14, 216
  settlement by Sir George Wingate,
    33, 61, 95, 134–8, 144–5,
    148–9, 151–4, 156, 218
  suspension and remission of, 12,
    17, 36, 66, 128–9, 132, 138,
    140, 146–7, 157–67 (**158, 160**),
    176, 194, 212–14, 219

Index   267

land revenue – *continued*
  zemindari system of, 113, 134,
    136–7, 147
  see also survey department
livestock, 32–3, 35–6 (**36**), 43, 45, 48,
  66, 74, 78, 95, 101–3, 111,
  120–1, 168, 199–202, 208
loans, *see* credit; *takavi* loans
Lytton, Lord, Viceroy, 132, 173–4, 201

markets, *see under* foodgrain markets;
  cotton, markets in; labour
  markets; credit, markets in
Marwaris, 17, 23, 38–9, 79, 92–114,
  117, 124, 126, 215
  see also credit; *sowcars*
Marx, Karl, 60, 99–100, 232 (n84)
McAlpin, Michelle, 4, 7, 33, 118,
  120–3, 136, 138, 218, 222 (n8),
  223 (n16), 234 (n131), 255 (n2)
migration, *see* coping strategies;
  Nizam of Hyderabad's
  Dominions
moneylending, *see* credit

Nasik district, 39, 41, 63, 65, 116,
  143, 149, **207**
newspapers
  Bombay Gazette, 9, 192
  Dnyan Prakash, 192, 247 (n6),
    252 (n132)
  Indu Prakash, 199, 247 (n5, n6)
  Jame Jamsed, 137, 178, 248 (n14)
  Nagar Samachar, 15, 82
  Native Opinion, 248 (n16)
  Nyaya Sindhu, 181
  Rast Goftar, 184
  The Times, 124
  Times of India, 182, 252 (n132)
Nizam of Hyderabad's Dominions,
  15, 19, 29, 63, 75–6, 85, 112,
  185, 187, 200–1
  migration to and from
    Ahmednagar, 200–2 (**201**), 208

peasants
  autonomy, constraints on, 14,
    16–17, 59, 79–80, 89, 92–4,
    100, 128, 212–14, 220

big, absence of, 4, 17, 21, 34, 78–9,
  90, 101, 110, 214
capital, lack of, 9, 23–5, 32–4, 78,
  89–90, 92, 109, 113, 135, 157,
  159, 168, 212–5
caution of, 34–5, 58–9, 75,
  78–82, 116, 139, 208, 213,
  215–16
cultivation strategies of, 3–4, 21,
  32–5, 38, 43–4, 59, 79–80, 114,
  139, 185
protests by, 13, 92, 102, 104–8,
  136–7, 141, 199, 212, 214,
  220–1
relationship to colonial state, 1,
  5–7, 9–10, 13–14, 17, 19–23,
  28–35, 41–51, 54–6, 58–9, 66,
  73, 80–1, 88–94, 97–9,
  102–204, 209–21
relationship to moneylenders, 5,
  17, 21, 71, 79–81, 89, 92–115,
  124, 212–15, 220
response to famine, 17, 170,
  195–204, 209, 216–18
vulnerability of, 2–4, 8–9, 13, 16,
  24, 31–2, 42, 45–8, 66, 75,
  80–1, 89–90, 93–5, 101–3,
  107–14, 118, 122, 126–30,
  135–204, 210–21
Pedder, Sir William, 23–4, 97, 108,
  112, 119, 130, 151, 233 (n92)
Poona district, 4–5, 19, 27, 35, 39,
  60, 63, 65, 92, 96, 105, 124,
  141, 143, 149, 151, 155, 163,
  187, **207**
Poona Sarvajanik Sabha, 14–15,
  124–6, 141, 147, 181–2, 192,
  196, 220, 248 (n8)

railways, *see* transport
rainfall, unreliability of, 2–3, 5, 7–8,
  17, 22, 34, 39–48 (**40**, **43**), 51,
  53, 56, 66, 80, 129, 139, 146,
  164–5, 167, 170, 208, 213,
  215, 217
relief wages, 12, 87–8, 181, 184,
  186–95 (**189**), **194**, 195–9,
  209–10
  see also Temple wage

Index

relief works, 2–3, 6, 63, 86–8, 164, 168–70, 217–18
  conditions on, 18, 181–3, 186–7, 190–1, 198–9, 205, 209–11
  costs of, 18, 53, 77, 170, 174–9 (**177**), 183–4, 211, 219
  distinction of civil agency and PWD, 77, 175–6, 180–3, 185–8, 191–5, 198, 203–4
  extent of exclusion of peasants from, 9, 13, 18, 56, 179, 195–6, 202–4, 206, 209–12, 216, 218–19
  tests of eligibility for, 81, 170, 177, 181–2, 184–98, 203–4, 207, 209–11, 220
  see also Government of Bombay – famine policy; gratuitous relief
riots, see Deccan riots; peasants, protest by
roads, see transport
Robertson, E. P., Revenue Commissioner, 13, 30, 45–7, 49–51, 53, 74–5, 77, 88, 102, 110–11, 118, 121, 123, 140, 143, 146, 156, 169, 175, 178, 180, 182, 184, 191, 194–5, 204, 209

sahukars, see sowcars
Salisbury, Lord, Secretary of State for India, 85, 130, 160, 173–4, 229 (n158)
Satara district, 4, **27**, 39, **60**, 79, 101, 110, 155, **207**
Sen, Amartya, 1, 42, 66, 83, 168, 204, 213, 223 (n9), 224 (n23)
  see also entitlements
Sholapur district, 4, **27**, 39, **60**, 110, 149, 155, 163, 169, 196–7, 200, **201**, 205, **207**
sowcars, 16, 23, 33, 59, 71, 79–81, 92–110, 112–17, 121, 123–6, 132, 151, 215
  see also credit
state
  hierarchy, 1, 10–13, 82, 88, 121, 129, 132–3, 147–55, 159, 165–7, 169–70, 172–9, 182, 185, 192–5, 197, 204, 210, 216, 219–21

policy-making, 1–2, 9–15, 21, 55, 72, 82, 87–8, 105, 126–7, 153–61, 165–70, 172–8, 182–92, 195–7, 204, 208–10, 212–13, 215, 219–21
role of individuals within, 10–14, 167, 169, 174–6, 182–4, 194–5, 210–11, 219–21
see also under particular governments and departments
statistics, unreliability of, 15–16, 20–2, 25–27, 30, 35–6, 40, 46, 49, 60–2, 73–4, 162, 204–6
Strachey, Sir John, Finance Member, Government of India (1873–81), 10, 12, 54, 82, 114–15, 119, 132–3, 159–60, 173–4, 178–9, 195, 215, 219
Strachey, Sir Richard, Chairman, Famine Commission (1880), 39, 172–4, 178
survey department, (Government of Bombay), 11, 16, 51–2, 68, 111, 114, 129, 133–4, 145–8, 150–4, 156, 167, 217
survival strategies, see coping strategies

takavi loans, 7, 17, 49, 93, 114–24 (**120**), 126, 152, 217–18
taxation, see land revenue
Temple, Sir Richard
  campaign in Bihar (1874), 82, 172–3, 177–9, 194, 203, 205, 209, 215
  control of famine expenditure in Bombay, 52, 120, 129, 160–7, 174, 176–80 (**177**), 187, 191–4, 208–10, 215–17
  as famine envoy to Madras, 12, 82, 160, 174
  as Finance Member, Government of India, 132
  mortality estimates of, 204–6, 208
  views on famine, prevention and relief, 41–2, 54, 72, 77, 86–9, 92, 172, 179, 185, 187, 195–6, 201–3, 206, 209, 216, 250 (n92)

Temple, Sir Richard – *continued*
   views on hierarchy, 12, 169, 174,
      194, 197
   views on land revenue system,
      143–4, 160–2
   views on peasants, 9, 92, 120, 132,
      179, 195–6, 200–3, 209, 216
Temple wage, 12–13, 170, 187–97,
   199, 203–4, 210
   refusal to accept, 14, 170, 179,
      192, 196–8, 209, 220
   see also relief wages
traders, 23–4, 57, 59, 65, 71, 75–6,
   79, 82–90, 131–2, 191, 209

transport, 4, 17, 58, 70–8, 84, 86,
   88–9, 136
   carts, **36**, 63, **73–74**
   railway, Dhond-Manmad,
      38, 47, 49, 63, 71–8,
      86–90, 145, 148, 174,
      176–7, 179–83, 199,
      201, 203, 207, 217
   roads, 71–3, 75–8, 145, 184,
      203

Wedderburn, Sir William,
   Sub-Judge, Ahmednagar, 123–6,
   141, 146–8